MEASURING RACIAL DISCRIMINATION

Panel on Methods for Assessing Discrimination

Rebecca M. Blank, Marilyn Dabady, and Constance F. Citro, Editors

Committee on National Statistics
Division of Behavioral and Social Sciences and Education

NATIONAL RESEARCH COUNCIL
OF THE NATIONAL ACADEMIES

THE NATIONAL ACADEMIES PRESS
Washington, D.C.
www.nap.edu

THE NATIONAL ACADEMIES PRESS 500 Fifth Street, N.W. Washington, DC 20001

NOTICE: The project that is the subject of this report was approved by the Governing Board of the National Research Council, whose members are drawn from the councils of the National Academy of Sciences, the National Academy of Engineering, and the Institute of Medicine. The members of the committee responsible for the report were chosen for their special competences and with regard for appropriate balance.

This study was supported by Contract/Grant Nos. 1000-0967, 40000660, 43-3AEP-0-80090, and ED-00-PO-4829 between the National Academy of Sciences and the Ford Foundation, the Andrew W. Mellon Foundation, the U.S. Department of Agriculture, and the U.S. Department of Education, respectively. Any opinions, findings, conclusions, or recommendations expressed in this publication are those of the author(s) and do not necessarily reflect the views of the organizations or agencies that provided support for the project.

Library of Congress Cataloging-in-Publication Data

Measuring racial discrimination / Panel on Methods for Assessing Discrimination, Committee on National Statistics, Division of Behavioral and Social Sciences and Education ; Rebecca M. Blank, Marilyn Dabady, and Constance F. Citro, editors.
p. cm.
Includes bibliographical references and index.
ISBN 0-309-09126-8 (hardcover) — ISBN 0-309-53083-0 (pdf)
1. Race discrimination—Research—Statistical methods. I. Blank, Rebecca M. II. Dabady, Marilyn. III. Citro, Constance F. (Constance Forbes), 1942- IV. National Research Council (U.S.). Panel on Methods for Assessing Discrimination.
HT1523.M43 2004
305.8—dc22
2004002885

Additional copies of this report are available from the National Academies Press, 500 Fifth Street, N.W., Lockbox 285, Washington, DC 20055; (800) 624-6242 or (202) 334-3313 (in the Washington metropolitan area); Internet, http://www.nap.edu.

Printed in the United States of America.

Suggested citation: National Research Council. (2004). *Measuring Racial Discrimination*. Panel on Methods for Assessing Discrimination. Rebecca M. Blank, Marilyn Dabady, and Constance F. Citro, Editors. Committee on National Statistics, Division of Behavioral and Social Sciences and Education. Washington, DC: The National Academies Press.

THE NATIONAL ACADEMIES

Advisers to the Nation on Science, Engineering, and Medicine

The **National Academy of Sciences** is a private, nonprofit, self-perpetuating society of distinguished scholars engaged in scientific and engineering research, dedicated to the furtherance of science and technology and to their use for the general welfare. Upon the authority of the charter granted to it by the Congress in 1863, the Academy has a mandate that requires it to advise the federal government on scientific and technical matters. Dr. Bruce M. Alberts is president of the National Academy of Sciences.

The **National Academy of Engineering** was established in 1964, under the charter of the National Academy of Sciences, as a parallel organization of outstanding engineers. It is autonomous in its administration and in the selection of its members, sharing with the National Academy of Sciences the responsibility for advising the federal government. The National Academy of Engineering also sponsors engineering programs aimed at meeting national needs, encourages education and research, and recognizes the superior achievements of engineers. Dr. Wm. A. Wulf is president of the National Academy of Engineering.

The **Institute of Medicine** was established in 1970 by the National Academy of Sciences to secure the services of eminent members of appropriate professions in the examination of policy matters pertaining to the health of the public. The Institute acts under the responsibility given to the National Academy of Sciences by its congressional charter to be an adviser to the federal government and, upon its own initiative, to identify issues of medical care, research, and education. Dr. Harvey V. Fineberg is president of the Institute of Medicine.

The **National Research Council** was organized by the National Academy of Sciences in 1916 to associate the broad community of science and technology with the Academy's purposes of furthering knowledge and advising the federal government. Functioning in accordance with general policies determined by the Academy, the Council has become the principal operating agency of both the National Academy of Sciences and the National Academy of Engineering in providing services to the government, the public, and the scientific and engineering communities. The Council is administered jointly by both Academies and the Institute of Medicine. Dr. Bruce M. Alberts and Dr. Wm. A. Wulf are chair and vice chair, respectively, of the National Research Council.

www.national-academies.org

PANEL ON METHODS FOR ASSESSING DISCRIMINATION

REBECCA M. BLANK (*Chair*), Gerald R. Ford School of Public Policy, University of Michigan
JOSEPH G. ALTONJI, Department of Economics, Yale University
ALFRED BLUMSTEIN, H. John Heinz III School of Public Policy and Management, Carnegie Mellon University
LAWRENCE BOBO,* Department of Sociology and Afro-American Studies, Harvard University
JOHN J. DONOHUE III, Stanford University School of Law
ROBERTO FERNANDEZ, Massachusetts Institute of Technology Sloan School of Management
STEPHEN E. FIENBERG, Department of Statistics, Carnegie Mellon University
SUSAN T. FISKE, Department of Psychology, Princeton University
GLENN C. LOURY, Institute on Race and Social Division, Boston University
SAMUEL R. LUCAS, Department of Sociology, University of California, Berkeley
DOUGLAS S. MASSEY, Woodrow Wilson School of Public and International Affairs, Princeton University
JANET L. NORWOOD, Chevy Chase, MD
JOHN E. ROLPH, Marshall School of Business, University of Southern California

MARILYN DABADY, *Study Director*
CONSTANCE F. CITRO, *Senior Program Officer*
MARISA A. GERSTEIN, *Research Assistant*
AGNES GASKIN, *Senior Project Assistant*
MARIA ALEJANDRO, *Project Assistant* through November 2002
DANELLE J. DESSAINT, *Senior Project Assistant* through July 2002

*Served on the panel until September 25, 2002.

Contents

Tables and Figures

TABLES

FIGURES

Preface

In the United States, large differences among racial and ethnic groups characterize many areas of social, economic, and political life, including such domains as the criminal justice system, education, employment, health care, and housing. For example, racial differences—which generally disadvantage minorities—exist in arrest and incarceration rates, earnings, income and wealth, levels of educational attainment, health status and health outcomes, and mortgage lending and homeownership. There are many possible explanations for such differences; one explanation may be the persistence of behaviors and processes of discrimination against minorities.

In this context, the Committee on National Statistics convened the Panel on Methods for Assessing Discrimination in 2001 to define racial discrimination; review and critique existing methods used to measure such discrimination and identify new approaches; and make recommendations regarding the best of these methods, as well as promising areas for future research. Because of wide interest in this topic, several funding agencies sponsored our study: the Ford Foundation, the Andrew W. Mellon Foundation, the U.S. Department of Agriculture, and the U.S. Department of Education.

The work of this panel is a direct outgrowth of the project that resulted in the two-volume report *America Becoming: Racial Trends and Their Consequences* (National Research Council, 2001a). Several of the panel members who were involved in producing these volumes held conversations around the question "What do we need to know to understand more about the role of race in American society?" At least one answer was

"We need better methods to identify and understand the effects of race-based discrimination."

The panel comprised a diverse group of experts in the fields of criminal justice, law, economics, psychology, public policy, sociology, and statistics. This diversity added a great deal to the creative debates among the panel members but also added to the difficulties in writing this report. It took time to develop a language and an intellectual framework with which we were all comfortable. In our report, we provide an extended discussion of definitions of discrimination and race, consolidating many aspects of a large social science literature on these topics. We also discuss various approaches to modeling and measuring discrimination in different fields. The interdisciplinary and diverse nature of the panel helped broaden these discussions, and we hope that our presentation of the definitional issues provides insight to those interested in the conceptualization of discrimination, just as we hope that our discussion of the methodological issues introduces new ideas to those engaged in measuring discrimination.

The breadth and complexity of the topic of discrimination and its effects posed a challenge for maintaining a tight focus on our charge, which was to *define* discrimination and review *methods* for measuring it. To keep to that charge, we spend no time discussing policies intended to alleviate discrimination (such as affirmative action or programs to build recruitment pools). We acknowledge, however, that the panel members have diverse opinions about appropriate policy options to address problems of discrimination, and inevitably our debates over policy issues at times crept into our debates over methodological issues.

Because of the charge and constraints on our time and resources, we focus our analysis on racial discrimination, particularly discrimination against African Americans, for which there is a very large literature. We do not address discrimination on the basis of nonracial factors, such as gender or age, nor do we discuss so-called reverse discrimination. Under the rubric of racial discrimination, we do include discrimination against ethnic groups, particularly Hispanics. The reasons have to do with the discrimination that has affected them coupled with the blurred nature of the definition of race and ethnicity for many Hispanics.

All of the panel members recognize the difficulties in defining racial discrimination in a clear way and in finding credible ways to measure it. There are different types of discrimination, different venues in which it can occur, and different ways in which it can have an effect. This report cannot address all of these topics comprehensively, but we have attempted to focus on at least some of the more important definitional and measurement problems. The measurement issues we address are relevant for understanding and measuring other types of discrimination. Despite the difficulty of our task, the panel members are all persuaded that accurate methods to identify and measure discrimination are highly important, and as scholars and researchers, we were committed to carrying out our charge in the best way possible.

I want to thank the people who have been important in making this report possible. Marilyn Dabady served as the study director for the report and devoted long hours and tireless effort to its production. The report could not have been written without her expertise and assistance. Senior program officer Constance Citro provided extremely helpful editing and writing assistance. Other staff members who contributed to the report in important ways were Seth Hauser and Michael Cohen. Danelle Dessaint and Agnes Gaskin, the panel's senior project assistants, provided outstanding assistance in organizing meetings, arranging travel, and preparing the final report. We are also grateful to Marisa Gerstein, who provided valuable research assistance to the panel.

Senior staff members Michael Feuer, Andrew White, Faith Mitchell, and Eugenia Grohman all provided useful advice to the panel as its deliberations proceeded. Our thanks to Rona Briere and Elaine McGarraugh for their careful editing of the report. Of course, we are grateful as well to our funders who made our work possible: the Ford Foundation, the Andrew W. Mellon Foundation, the U.S. Department of Education, and the U.S. Department of Agriculture. We especially want to thank Joseph Meisel (Program Officer for Higher Education, Mellon Foundation), Alan Jenkins (Director, Human Rights, Ford Foundation) and Sara Rios (Program Officer, Ford Foundation), Marilyn Seastrom (Chief Statistician, National Center for Education Statistics, U.S. Department of Education), Susan Offutt (Administrator, Economic Research Service, U.S. Department of Agriculture), and their colleagues for their continued interest in this project.

A number of outside experts contributed valuable information for this study. Those who wrote commissioned papers for the panel included George Farkas, Pennsylvania State University; Harry Holzer, Georgetown University; Jens Ludwig, Georgetown University; Roslyn Mickelson, University of North Carolina-Charlotte; Robert Nelson and Eric Bennett, Northwestern University; Stephen Ross, University of Connecticut; James Ryan, Yale University; Thomas Smith, University of Chicago, National Opinion Research Center; and John Yinger, Syracuse University. Others testified to the panel on important issues. They included David Harris, University of Michigan; Rebecca Fitch, Office for Civil Rights, U.S. Department of Education; Richard Foster, Office for Civil Rights, U.S. Department of Education; Susan Offutt, Economic Research Service, U.S. Department of Agriculture; Todd Richardson, U.S. Department of Housing and Urban Development; Dan Sutherland, Chief of Staff, Office for Civil Rights, U.S. Department of

Education; Clyde Tucker, Senior Statistician, Bureau of Labor Statistics; Katherine Wallman, Chief Statistician, U.S. Office of Management and Budget; and Matthew Zingraff, North Carolina State University.

The panel also appreciates the useful assistance and insight of many colleagues during its deliberations. They include Ronald Ferguson, Harvard University; Joan First, National Coalition of Advocates for Students; Willis Hawley, University of Maryland; Judith Hellerstein, University of Maryland; John Kain, University of Texas-Dallas; Valerie Lee, University of Michigan; Jeanette Lim, Office for Civil Rights, U.S. Department of Education; Michael Rebell, Campaign for Fiscal Equality, Inc.; Francine Blau, Cornell University; David Card, University of California-Berkeley; Lindsay Chase-Landsdale, Northwestern University; Celina M. Chatman, University of Michigan; George Galster, Wayne State University; Robert Hauser, University of Wisconsin-Madison; Christopher Jencks, Harvard University; Nancy Krieger, Harvard University; Susan Murphy, University of Michigan; and Christopher Winship, Harvard University.

This report has been reviewed in draft form by individuals chosen for their diverse perspectives and technical expertise, in accordance with procedures approved by the National Research Council's Report Review Committee. The purpose of this independent review is to provide candid and critical comments that will assist the institution in making the published report as sound as possible and to ensure that the report meets institutional standards for objectivity, evidence, and responsiveness to the study charge. The review comments and draft manuscript remain confidential to protect the integrity of the deliberative process. We wish to thank the following individuals for their review of this report: John C. Bailar III, Department of Health Studies (emeritus), University of Chicago; Francine D. Blau, School of Industrial and Labor Relations, Cornell University; William Darity, Jr., Department of Economics, University of North Carolina; Christopher Edley, Law School, Harvard University; Richard A. Epstein, Law School, University of Chicago; Paul Holland, Educational Testing Service, Princeton, New Jersey; James M. Jones, Department of Psychology, University of Delaware; Shelly Lundberg, Center for Research on Families, University of Washington; Ewart A.C. Thomas, Department of Psychology, Stanford University; Larry Wasserman, Department of Statistics, Carnegie Mellon University; and David R. Williams, Institute for Social Research, University of Michigan.

Although the reviewers listed above provided many constructive comments and suggestions, they were not asked to endorse the report's conclusions or recommendations nor did they see the final draft of the report before its release. The review of this report was overseen by Cora B. Marrett, Senior Vice President for Academic Affairs, University of Wisconsin System

Administration, and Lyle V. Jones, L.L. Thurstone Psychometric Laboratory, University of North Carolina. Appointed by the National Research Council, they were responsible for making certain that an independent examination of this report was carried out in accordance with institutional procedures and that all review comments were carefully considered. Responsibility for the final content of this report rests entirely with the authoring committee and the institution.

Finally, and most important, I thank the panel members themselves. Our discussions have been challenging, contentious, humorous, frustrating, enjoyable, and always intellectually stimulating. Each panel member contributed much time and effort to intellectually shaping, writing, editing, and critiquing this report. I believe the final product reflects the level of interest, concern, and commitment every panel member brought to the table.

Rebecca M. Blank, *Chair*
Panel on Methods for Assessing Discrimination

Executive Summary

Many racial and ethnic groups in the United States, including blacks, Hispanics, Asians, American Indians, and others, have historically faced severe discrimination—pervasive and open denial of civil, social, political, educational, and economic opportunities. Today, large differences in outcomes among racial and ethnic groups continue to exist in employment, income and wealth, housing, education, criminal justice, health, and other areas. Although many factors may contribute to such differences, their size and extent suggest that various forms of discriminatory treatment persist in U.S. society and serve to undercut the achievement of equal opportunity.

In these circumstances, it is critically important to identify where racial discrimination occurs and to measure the extent to which discrimination may contribute to racial and ethnic disparities. The Committee on National Statistics convened a panel of scholars to consider the definition of racial discrimination, assess current methodologies for measuring it, identify new approaches, and make recommendations about the best broad methodological approaches. Specifically, this panel was asked to carry out the following tasks:

1. Give the policy and scholarly communities new tools for assessing the extent to which discrimination continues to undermine the achievement of equal opportunity by suggesting additional means for measuring discrimination that can be applied not only to the racial question but in other important social arenas as well.

2. Conduct a thorough evaluation of current methodologies for measuring discrimination in a wide range of circumstances where it may occur.

3. Consider how analyses of data from other sources could contribute to findings from research experimentation, such as the U.S. Department of Housing and Urban Development paired tests.

4. Recommend further research as well as the development of data to complement research studies.

DEFINING RACE

There is no single concept of race. Rather, race is a complex concept, best viewed for social science purposes as a subjective social construct based on observed or ascribed characteristics that have acquired socially significant meaning. In the United States, ways in which different populations think about their own and others' racial status have changed over time in response to changing patterns of immigration, changing social and economic situations, and changing societal norms and government policies. In the late nineteenth and early twentieth centuries, for example, some European Americans, such as Italians and Eastern European Jews, were regarded as distinct racial groups. Although these distinctions are no longer sanctioned by the U.S. government, some segments of the population may still act in ways that are consistent with such distinctions. For certain populations and in some situations, race may be difficult to define consistently; for example, many Hispanics consider themselves to be part of a distinct racial group, but many others hold no such perception. Because concepts of race and ethnicity are not clearly defined for many Hispanics and because of the discrimination they have faced, we include Hispanics, along with specific racial groups, in our discussion of racial discrimination.

The ambiguity involved in defining race has implications for how data on race are collected. The official federal government standards for data on race and ethnicity currently identify five major racial groups (black or African American, American Indian or Alaska Native, Asian, Native Hawaiian or other Pacific Islander, and white) and one ethnic group (Hispanic) that may be of any race. These categories are used by federal program and statistical agencies to collect data through self-reports (preferably) or by assigning individuals to one or more categories. The federal racial categories have changed over time, in part reflecting the changing conception of race in the United States. The government standards are not always consistent with scholarly concepts of race or with concepts held by individuals and groups; as a result, it may be difficult to obtain data on race and ethnicity that are comparable over time or across different surveys and administrative records. Comparability may also be affected by differences in the data collection methods used. Yet given the salience of race in so many aspects of social, political, and economic life, it is important to continue collecting these data.

> **Conclusion:** *For the purpose of understanding and measuring racial discrimination, race should be viewed as a social construct that evolves over time. Despite measurement problems, data on race and ethnicity are necessary for monitoring and understanding evolving differences and trends in outcomes among groups in the U.S. population. (from Chapters 2 and 10)*

> **Recommendation: The federal government and, as appropriate, state and local governments should continue to collect data on race and ethnicity. Federal standards for racial categories should be responsive to changing concepts of race among groups in the U.S. population. Any resulting modifications to the standards should be implemented in ways that facilitate comparisons over time to the extent possible. (Recommendation 10.1)[1]**

> **Recommendation: Data collectors, researchers, and others should be cognizant of the effects of measurement methods on reporting of race and ethnicity, which may affect the comparability of data for analysis:**
>
> - **To facilitate understanding of reporting effects and to develop good measurement practices for data on race, federal agencies should seek ways to test the effects of such factors as data collection mode (e.g., telephone, personal interview), location (e.g., home, workplace), respondent (e.g., self, parent, employer, teacher), and question wording and ordering. Agencies should also collect and analyze longitudinal data to measure how reported perceptions of racial identification change over time for different groups (e.g., Hispanics and those of mixed race).**
>
> - **Because measurement of race can vary with the method used, reports on race should to the extent practical use multiple measurement methods and assess the variation in results across the methods. (Recommendation 10.2)**

---

[1]For ease of reference, the panel's recommendations are numbered according to the chapter of the report in which they appear. For example, Recommendation 10.1 is the first recommendation presented in Chapter 10.

DEFINING RACIAL DISCRIMINATION

This report adopts a social science definition of racial discrimination that has two components:

(1) *differential treatment on the basis of race* that disadvantages a racial group and

(2) *treatment on the basis of inadequately justified factors other than race* that disadvantages a racial group (differential effect).

In this report, we focus on discrimination against disadvantaged racial minorities. The two components of our definition—differential treatment and differential effect discrimination—are related to but broader than the standards embodied in case law in the U.S. legal system, which are *disparate treatment* and *disparate impact discrimination.* An example of potentially unlawful disparate treatment discrimination would be when an individual is not hired for a job because of his or her race. An example of potentially unlawful disparate impact discrimination would be when an employer uses a test in selecting job applicants that is not a good predictor of performance on the job and results in proportionately fewer job offers being extended to members of disadvantaged racial groups compared with whites.[2]

Because our intention in this report is to provide guidance to social science researchers interested in measuring discrimination, both components of our definition include a range of behaviors and processes that are not explicitly unlawful or easily measured. For example, many governmental actions that might fall within the legal definition of disparate impact discrimination would not be unlawful because the Supreme Court has interpreted the constitutional prohibition on denials of equal protection by government agencies to bar only cases of intentional discrimination—that is, disparate treatment discrimination. As a second example, discrimination would occur under our definition when interviewers of job applicants more frequently adopt behaviors (e.g., interrupting, asking fewer questions, using a hectoring tone) that result in poorer communication with and performance by disadvantaged minority applicants compared with other applicants. Even if such behaviors became the subject of a legal challenge, the difficulties in measurement and proof would likely mean that such behav-

[2]We use the term *disadvantaged racial groups* interchangeably with *minority groups* to refer to blacks, American Indians or Alaska Natives, Asians, Native Hawaiians or other Pacific Islanders, and, in some cases, Hispanics. Members of these groups have more often been discriminated against in various social and economic arenas.

iors would not be effectively constrained by law. Measuring them is important, however, to understand ways in which subtle forms of discrimination may affect important social and economic outcomes.

MEASURING RACIAL DISCRIMINATION

That racial disparities exist in a wide range of social and economic outcomes is not in question: They can be seen in higher rates of poverty, unemployment, and residential segregation and in lower levels of education and wealth accumulation for some racial groups compared with others. Large and persistent outcome differences, however, do not themselves provide direct evidence of the presence or magnitude of racial discrimination in any particular domain. Differential outcomes may indicate that discrimination is occurring, that the historical effects of racial exclusion and discrimination (cumulative disadvantage) continue to influence current outcomes, that other factors are at work, or that some combination of current and past discrimination and other factors is operating.

The panel evaluated four major methods used across different social and behavioral science disciplines to measure racial discrimination: laboratory experiments, field experiments, analysis of observational data and natural experiments, and analysis of survey and administrative record reports. Each method has strengths and weaknesses, particularly for drawing a causal inference that an adverse outcome is the result of race-based discriminatory behavior.

Because discriminatory behavior is rarely observed directly, researchers must infer its presence by trying to determine whether an observed adverse outcome for an individual would have been different had the individual been of a different race. In other words, researchers attempt to answer the following counterfactual question: What would have happened to a nonwhite individual if he or she had been white? Understanding the extent to which any study succeeds in answering that question requires rigorously assessing the logic and assumptions underlying the causal inferences drawn by the researchers. As was true in determining that smoking causes lung cancer, using a variety of methods implemented in a variety of settings is likely to be most helpful in measuring discrimination.

> **Conclusion:** *No single approach to measuring racial discrimination allows researchers to address all the important measurement issues or to answer all the questions of interest. Consistent patterns of results across studies and different approaches tend to provide the strongest argument. Public and private agencies—including the National Science Foundation, the National Institutes of Health, and private founda-*

tions—and the research community should embrace a multidisciplinary, multimethod approach to the measurement of racial discrimination and seek improvements in all major methods employed. (from Chapter 5)

Laboratory Experiments

Classically, laboratory experimentation in which a stimulus can be administered to research participants in a controlled environment and in which participants can be randomly assigned to an experimental condition or another (e.g., control) condition provides the best approach for inferring causation between a stimulus and a response. Such experiments come closest to addressing the above counterfactual question.

Laboratory experiments have uncovered many subtle yet powerful psychological mechanisms through which racial bias exists. Yet regardless of how well designed and executed they are, laboratory experiments cannot by themselves directly address how much race-based discrimination against disadvantaged groups contributes to adverse outcomes for those groups in society at large.

The major contributions of laboratory experiments are to identify those situations in which discriminatory attitudes and behaviors are more or less likely to occur, as well as the characteristics of people who are more or less likely to exhibit discriminatory attitudes and behaviors, and to provide models of people's mental processes that may lead to racial discrimination. Such experiments can usefully suggest hypotheses to be tested with other methodologies and real-world data.

Recommendation: To enhance the contribution of laboratory experiments to measuring racial discrimination, public and private funding agencies and researchers should give priority to the following:

- **Laboratory experiments that examine not only racially discriminatory attitudes but also discriminatory behavior. The results of such experiments could provide the theoretical basis for more accurate and complete statistical models of racial discrimination fit to observational data.**

- **Studies designed to test whether the results of laboratory experiments can be replicated in real-word settings with real-world data. Such studies can help establish the general applicability of laboratory findings. (Recommendation 6.1)**

Field Experiments

Large-scale experiments in the field rely on random assignment of subjects to one or more experimental treatments or to no treatment, so that researchers can determine whether an experimental treatment (the stimulus) causes an observed response. Such experiments take longer and are more complex to manage and more costly to conduct than laboratory experiments, and their results are more easily confounded by factors in the environment that the researchers cannot control. However, their results are more readily generalizable to the population at large.

The most significant use of field studies to study discrimination to date has been in the area of housing, specifically seeking new apartments or houses. The results of audit or paired-testing studies—in which otherwise comparable pairs of, say, a black person and a white person are sent separately to realty offices to seek an apartment or house—have been used to measure discrimination in specific housing markets. Audit studies have also been conducted on job seeking. It is likely that audit studies of racial discrimination in other domains (e.g., schooling and health care) could produce useful results as well, even though their use will undoubtedly present methodological challenges specific to each domain.

Recommendation: Nationwide field audit studies of racially based housing discrimination, such as those implemented by the U.S. Department of Housing and Urban Development in 1977, 1989, and 2000, provide valuable data and should be continued. (Recommendation 6.2)

Recommendation: Because properly designed and executed field audit studies can provide an important and useful means of measuring discrimination in various domains, public and private funding agencies should explore appropriately designed experiments for this purpose. (Recommendation 6.3)

Statistical Analysis of Observational Data and Natural Experiments

Observational studies are currently the primary tool through which researchers explore issues of racial disparity and discrimination in the real world. The standard way to explore the difference in an outcome between racial groups is to develop a regression model that includes a variable for race and variables for other relevant observed characteristics. The effect of the former variable on the outcome difference is identified as discrimination.

To support a causal inference from observational data, however, substantial prior knowledge about the mechanisms that generated the data must be available to justify the necessary assumptions. There are two particularly common problems involved in using standard multiple regression models to analyze observational data on outcome differences between race groups: *Omitted variables bias* occurs whenever a data set contains only a limited number of the characteristics that may reasonably factor into the process under study; *sample selection bias* occurs when the research systematically excludes subjects from the sample whose characteristics vary from those of the individuals represented in the data. Should either bias be present, it is difficult to draw causal inferences from the coefficient on race (or any other variable) in a regression model, as the race coefficient may overestimate or underestimate the effect labeled as discrimination.

Nationally representative data sets containing rich measures of the variables that are the most important determinants of such outcomes as education, labor market success, and health status can help in estimating and understanding the sources of racial differences in outcomes. Panel data, which include observations over time, are particularly valuable in this regard. There is also an important role for focused studies that target particular settings (e.g., a firm or a school), whereby it is possible to learn a great deal about how decisions are made and to collect most of the information on which decisions are based.

Evaluations of natural experiments are another way to exploit observational data in the measurement of racial discrimination. Such evaluations analyze data before and after enactment of a new law or some other change that forces a reduction in or the complete elimination of discrimination for some groups. Despite limitations, natural experiments provide useful data for measuring the extent of discrimination prior to a policy change and for groups not affected by the change.

Conclusion: *The statistical decomposition of racial gaps in social outcomes using multivariate regression and related techniques is a valuable tool for understanding the sources of racial differences. However, such decompositions using data sets with limited numbers of explanatory variables, such as the Current Population Survey or the decennial census, do not accurately measure the portion of those differences that is due to current discrimination. Matching and related techniques provide a useful alternative to race gap decompositions based on multivariate regression in some circumstances. (from Chapter 7)*

Conclusion: *The use of statistical models, such as multiple regressions, to draw valid inferences about discriminatory behavior requires appropriate data and methods, coupled with a sufficient understanding* *of the process being studied to justify the necessary assumptions. (from Chapter 7)*

Recommendation: Public and private funding agencies should support focused studies of decision processes, such as the behavior of firms in hiring, training, and promoting employees. The results of such studies can guide the development of improved models and data for statistical analysis of differential outcomes for racial and ethnic groups in employment and other areas. (Recommendation 7.1)

Recommendation: Public agencies should assist in the evaluation of natural experiments by collecting data that can be used to evaluate the effect of antidiscrimination policy changes on groups covered by the changes as well as groups not covered. (Recommendation 7.2)

Indicators of Discrimination from Surveys and Administrative Records

Both self-reports of racial attitudes and perceived experiences of discrimination in surveys and reports of discriminatory events in administrative records can contribute to understanding the extent of racial discrimination. Survey data typically cannot directly measure the prevalence of actual discrimination as opposed to reports of perceived discrimination, but they can provide useful supporting evidence. Perceived discrimination may overreport or underreport discrimination assessed by other methods. As expressions of prejudice and discriminatory behavior change over time and become more subtle, new or revised survey questions on racial attitudes and perceived experiences of discrimination may be necessary. Longitudinal and repeated cross-sectional data, including continuous and new measures, are important to illuminate trends and changes in patterns of racially discriminatory attitudes and behaviors among and toward various groups. Such data are also vital for studies of cumulative disadvantage. Administrative reports of discrimination (e.g., equal employment opportunity complaints) may also be useful for research, although the lack of completeness and reliability of such reports can limit their usefulness.

Recommendation: To understand changes in racial attitudes and reported perceptions of discrimination over time, public and private funding agencies should continue to support the collection of rich survey data:

- **The General Social Survey, which since 1972 has been the leading source of repeated cross-sectional data on trends in racial attitudes and perceptions of racial discrimination, merits continued support**

for measurement of important dimensions of discrimination over time and among population groups.

- **Major longitudinal surveys, such as the Panel Study of Income Dynamics, the National Longitudinal Survey of Youth, and others, merit support as data sources for studies of cumulative disadvantage across time, domains, generations, and population groups. To further enhance their usefulness, questions on perceived experiences of racial discrimination and racial attitudes should be added to these surveys.**

- **Data collection sponsors should support research on question wording and survey design that can lead to improvements in survey-based measures relating to perceived experiences of racial discrimination. (Recommendation 8.1)**

Recommendation: Agencies that collect administrative record reports of racial discrimination should seek ways to allow researchers to use these data for analyzing discrimination where appropriate. They should also identify ways to improve the completeness, reliability, and usefulness of reports of particular types of discriminatory events for both administrative and research purposes. (Recommendation 8.2)

Racial Profiling as an Illustrative Example

To provide a specific example of an area for which research on discriminatory treatment is needed but difficult to carry out, we discuss methodological issues in profiling. Racial or ethnic profiling is a screening process in which some individuals in a population (e.g., automobile drivers or people boarding an airplane) are selected on the basis of their race or ethnicity (and, typically, other observable characteristics) and investigated to determine whether they have committed or intend to commit a criminal act (e.g., smuggle drugs or blow up an airplane) or other act of interest. This definition excludes cases of identified individuals for whom race or ethnicity is part of their individual description. Many recent public statements (e.g., those made by police officials and legislative bodies since 2001) have recognized the unacceptability of racial profiling in police work. Even when such profiling is not explicitly racial, to the extent that it relies on characteristics that are distributed differently for different racial groups, the result may be a racially disparate impact.

Inferring the presence of discriminatory racial profiling from data on disparate outcomes is difficult for the same reasons that it is difficult to infer causation from any statistical model with observational data. We explore specific methodological concerns for improving the estimation of outcome rates (e.g., traffic stops for whites and minorities) and developing good statistical models for determining the contribution of discriminatory profiling as compared with other factors to differences in rates. Because of renewed interest in the United States in the use of profiling to identify and apprehend potential terrorists before they commit violent acts, we also examine briefly the challenges of identifying screening factors that could potentially select would-be terrorists with a significantly higher probability than purely random selection, as well as issues that must factor into the public debate if race or ethnicity (or factors that correlate highly with race or ethnicity) are considered as potential screening factors.

CUMULATIVE DISCRIMINATION

Much of the discussion about the presence of racial discrimination and the effects of antidiscrimination policies assumes discrimination to be a phenomenon that occurs at one point in time in a particular process or stage of a particular domain (e.g., initial hires by employers). This episodic view of discrimination is likely inadequate. Discrimination may well have cumulative effects, and it is therefore better viewed as a dynamic process that functions throughout the stages within a domain, across domains, across individual lifetimes, and even across generations. For example, discrimination involving teachers' expectations during schooling may affect students' later educational experiences or job opportunities; likewise, discrimination against prior generations may diminish opportunities for present generations even in the absence of current discriminatory practices.

Several theories of the processes by which discrimination may have cumulative effects have been developed, including (1) life-course theory of cumulative disadvantage in criminal justice research, which posits that such behavior as juvenile delinquency can affect certain social outcomes, such as failure in school or poor job stability, and thereby facilitate criminal behavior as an adult; (2) ecosocial theory in public health research (similar to the life-course concept), in which health status at a given age for a given birth cohort reflects not only current conditions but also prior living circumstances from conception onward; and (3) feedback models in labor market research. In such a model, for example, people who anticipate lower future returns to skills—possibly as a result of racial discrimination—might invest less in acquiring those skills. In turn, lower investment could perpetuate prejudice, limit opportunities, and sustain racial disparities in the labor market.

Only very limited research has been conducted, however, to test empirically the various theories of cumulative disadvantage and to measure the importance of cumulative effects over time and across domains. Longitudinal data are a necessity for such research, as are methods for credibly identifying initial and subsequent incidents of discrimination.

Conclusion: *Measures of discrimination from one point in time and in one domain may be insufficient to identify the overall impact of discrimination on individuals. Further research is needed to model and analyze longitudinal and other data and to study how effects of discrimination may accumulate across domains and over time in ways that perpetuate racial inequality. (from Chapter 11)*

Recommendation: Major longitudinal surveys, such as the Panel Study of Income Dynamics, the National Longitudinal Survey of Youth, and others, merit support as data sources for studies of cumulative disadvantage across time, domains, generations, and population groups. Furthermore, consideration should be given to incorporating into these surveys additional variables or special topical modules that might enhance the utility of the data for studying the long-term effects of discrimination. Consideration should also be given to including questions in new longitudinal surveys that would help researchers identify experiences of discrimination and their effects. (Recommendation 11.1)

NEXT STEPS

Our report emphasizes the challenges of measuring racial discrimination in various social and economic domains. Establishing that discriminatory treatment or impact has occurred and measuring its effects on outcomes requires very careful analysis to rule out alternative explanatory factors. In some research to date, the data and analytical methods used are not sufficient to justify the assumptions of the underlying theoretical model. Moreover, many analyses never articulate an explicit model, which makes it difficult to judge the adequacy of the data and analysis to support the study findings.

Just because it is challenging to measure discrimination does not mean that sound, adequate research in this area is not possible. To the contrary, existing methods and data have produced useful results on particular types of discrimination in particular aspects of a domain or process. To make further progress, we believe it will be necessary for funding and program agencies to support research that cuts across disciplinary boundaries, makes use of multiple methods and types of data, and studies racial discrimination as a dynamic process. To be cost-effective, such research should be focused and designed to maximize the analytical value of existing bodies of knowledge and ongoing surveys and administrative records data collections.

Agencies with programmatic responsibilities (e.g., to monitor discrimination, investigate complaints, and operate programs that may be affected by the presence of discrimination and by antidiscrimination laws and regulations) will need to single out priority areas of concern and develop detailed research plans for them. This may require studies of key decision-making processes, combined with theoretical models of the ways in which discrimination might occur. For this purpose, the existing literature of laboratory experiments about the kinds of situations in which discriminatory attitudes are most likely to lead to race-based discriminatory treatment should be reviewed and additional experiments commissioned, if the laboratory results are not sufficiently revealing about the decision processes of interest (e.g., employer decisions about job training and promotion, to take a labor market example). In turn, experimental results can help guide focused case studies of decision processes that may be needed to provide the requisite depth of understanding to permit subsequent statistical analysis with appropriate data and methods. To facilitate data availability and use, program agencies can not only support the addition of relevant questions to ongoing cross-sectional and longitudinal surveys but also work to improve the research potential of agency administrative records data.

Research agencies, both public and private, can best leverage their resources by addressing important areas of research on racial discrimination that are less apt to be considered by program agencies. In particular, they are better positioned to support innovative, cross-disciplinary, multimethod research on cumulative disadvantage. They can also usefully consider ways to augment ongoing and new panel surveys to provide relevant data for basic research on racial discrimination, particularly over long periods of time. The kinds of multifaceted studies that have been conducted in recent years of changes in the well-being of low-income populations following major changes in welfare policies may offer useful guidance for discrimination research, which could similarly make use of multiple data sources and perspectives from economics, psychology, ethnography, survey research, and other relevant disciplines. Such complex research will be difficult to conceptualize and carry out, but it offers the promise to expand knowledge about the role that current and past discrimination may play in shaping American society today.

1

Introduction

Most people would agree that equal opportunity to participate as a full and functioning member of society is important. Nonetheless, existing social and economic disparities among racial and ethnic groups suggest that our society has yet to achieve this goal. For instance, Hispanics have higher school dropout rates than other racial and ethnic groups (Hauser et al., 2002). The black–white wealth gap remains large (Conley, 1999; Oliver and Shapiro, 1995). Young Native Americans are incarcerated in federal prisons at higher rates than any other minority racial group (Smelser and Baltes, 2001; Weich and Angulo, 2002). And some Asian Americans, among other minority groups, have poorer access to health care services and treatments than whites (Institute of Medicine, 2003). Such racial disparities are pervasive and may be the result of racial prejudice and discrimination, as well as differences in socioeconomic status, differential access to opportunities, and institutional policies and practices.

Such racial disparities persist despite the many legal and social changes that have improved opportunities for minority racial and ethnic groups in the United States. Several factors may contribute to racial differences in outcomes, including differences in socioeconomic status, differential access to opportunities, and others. One factor that should be considered is the role of racial discrimination. Overt discrimination against African Americans and other minority groups characterized much of U.S. history; a question is whether and what types of discrimination continue to exist and their effects on differential outcomes.

Although researchers in specific disciplines have investigated discrimination in particular domains, there has been little effort to coordinate and expand such research in ways that could help to better understand and measure various kinds of racial and ethnic discrimination across domains and groups and over time. To address this problem, the Committee on National Statistics convened a panel of scholars in 2001 to consider the definition of racial discrimination, assess current methodologies for measuring it, identify new approaches, and make recommendations about the best broad methodological approaches. In particular, this panel was asked to conduct the following tasks:

1. Give the policy and scholarly communities new tools for assessing the extent to which discrimination continues to undermine the achievement of equal opportunity by suggesting additional means for measuring discrimination that can be applied not only to the racial question but in other important social arenas as well.
2. Conduct a thorough evaluation of current methodologies for measuring discrimination in a wide range of circumstances where it may occur.
3. Consider how analyses of data from other sources could contribute to findings from research experimentation, such as the U.S. Department of Housing and Urban Development paired tests.
4. Recommend further research as well as the development of data to complement research studies.

Although there is substantial direct empirical evidence for the prevalence of large disparities among racial and ethnic groups in various domains, it is often difficult to obtain direct evidence of whether and to what extent discrimination may be a contributing factor. Differential outcomes by race and ethnicity may or may not indicate discrimination. Examples of studies using methods that persuasively measure the presence or absence of discrimination are rare, and appropriate data for measurement are often unobtainable. As a result, there is little scholarly consensus about the extent and frequency of discrimination and how it relates to continuing disadvantages along racial and ethnic lines (Fix and Turner, 1998).

One reason it is difficult to assess discrimination is that changes have occurred in the nature of prejudiced attitudes and discriminatory behaviors. With the passage of the Civil Rights Act of 1964 and other laws that prohibit discrimination because of race in a variety of domains, overt discrimination is less often apparent. However, discrimination may persist in more subtle forms. Indeed, social psychological research suggests that relatively automatic and unexamined cognitive processes, of which the holder (and sometimes the target) may not be fully aware, can lead to discrimination (Devine, 1989; Fiske, 1998). These subtleties make defining and measuring discrimination more difficult.

STUDY APPROACH AND SCOPE

The panel's goal in this report is to review and comment on the methods used in various social scientific disciplines to identify types of racial discrimination and measure their effects. The report is designed to help social science researchers, policy analysts, federal agencies, and concerned observers better understand how to assess racial discrimination in different domains, drawing on different social science methods and data sources as appropriate. To approach this important but difficult task, the panel focused on defining relevant concepts, examining various methodological approaches and data sources, and considering directions for future research.

The purpose of this report is not to promote a single "right" way to measure discrimination. In some situations, one approach may be more easily implemented and more credible; in other situations, another approach may be more appropriate. Often, multiple approaches will be needed to provide credible evidence about the prevalence of discrimination in a domain. Thus, the panel attempts to identify the broad range of approaches for measuring discrimination and to provide a critical review of their relative credibility when applied in different situations. The panel develops a cross-disciplinary research and data collection agenda for action by public and private funding agencies and the research community.

The report makes no attempt to actually measure current or past levels of discrimination in any domain. Our purpose is not to report numbers or impacts but to provide guidance and encouragement to researchers and policy analysts as they work across domains to identify where discrimination may be present and what its effects may be.

In the first part of this report, the panel defines the concepts of race and racial discrimination from a social science perspective, which we believe is the appropriate perspective for research and policy analysis on discrimination. When referring to race in the report, the panel uses the categories established by the federal classification standards (U.S. Office of Management and Budget, 1997) to identify whites, blacks or African Americans, American Indians and Alaska Natives or Native Americans, Asians, and Native Hawaiians and other Pacific Islanders. According to these standards, Hispanics or Latinos are referred to as an *ethnic* group. Yet, although the panel was asked to consider *racial* discrimination, Hispanics (a rapidly growing ethnic population) also face discrimination. In addition, concepts of race and ethnicity are not clearly defined for many Hispanics, so for these two reasons our discussion often refers to Hispanics as well as to specific racial groups. Throughout the report, the term *disadvantaged racial group* is used to refer to groups in the United States (e.g., blacks) whose disadvantage can be linked historically to discriminatory practices and policies and who are, consequently, part of a legally protected class.

The panel is concerned with broad types of discriminatory behaviors and processes that have negative consequences for disadvantaged racial groups in various social and economic arenas. We draw on sociological, social psychological, and other literature to develop our definition of racial discrimination. We also discuss the conceptual possibility that discrimination may operate not just at one point in time and within one particular domain but at various points within and across multiple domains throughout the course of an individual's life. The panel acknowledges that the effect of such cumulative discrimination may not be easily identified or measured.

In interpreting that part of its charge to review measurement methods, the panel chose to address broad approaches that could be applied across domains, rather than making recommendations about specific approaches for particular domains. Therefore, although examples are used throughout the report to illustrate efforts to measure discrimination in particular circumstances, our main focus is on methods (e.g., experiments, observational studies, survey research) that can be used to study discrimination under many different circumstances.

The examples of disparities and discrimination measurement that we provide come from research in five domains: labor markets and employment, education, housing and mortgage lending, health care, and criminal justice. Although not the only domains of concern, these are key areas of social interaction for which discrimination can seriously limit life opportunities; these are also among the areas for which the federal government regularly collects administrative and survey data long used by researchers to study discrimination and discriminatory effects. We do not provide an exhaustive set of examples for each of these areas. Rather, a selected bibliography of important literature reviews, major reports, and other work on data collection and analytical methods used in each of these domains is provided at the end of this report.

Much of the discussion in this report on such topics as statistical inference, experimental design, and data quality is relatively technical in nature. Although sometimes dry, the import of this discussion should not be misunderstood by readers who are deeply concerned about the possible extent and continued effects of racial discrimination in American life. It was our shared concern about racial discrimination that drew each member of the panel into the in-depth discussions of measurement reflected in this report. Because we view racial discrimination as a crucial social issue, we believe it is essential to use the most credible and accurate measurement approaches.

In carrying out this study, the panel met and deliberated over a period of almost 2 years. We held meetings, invited speakers, and commissioned several papers (see Box 1-1); we requested input from prominent scholars on key issues; reviewed a large body of literature on salient aspects of the law and criminal justice, labor markets, housing markets, education, and

BOX 1-1
Papers Commissioned for This Study

Three papers were commissioned to inform the panel's work on this report. Smith (2002) reviews methods for measuring racial discrimination, focusing primarily on survey-based approaches. Ross and Yinger (2002) examine the use and quality of data on race collected for administrative purposes, as well as issues of comparability and interpretation that arise for both enforcement officials and scholars attempting to study discrimination. Finally, Nelson and Bennett (2003) review the courts' use of statistics to make decisions in cases alleging racial discrimination in employment. These papers are available directly from the authors.

The panel also commissioned several papers for a workshop on measuring racial disparities and discrimination in elementary and secondary education (see Appendix A). The purpose of the workshop was to expand and improve the statistical capability of the U.S. Department of Education and other federal agencies to measure and track discrimination. The four commissioned papers relating to measuring racial disparities and discrimination in education are published in *Teacher's College Record* (Farkas, 2003; Holzer and Ludwig, 2003; Mickelson, 2003; Ryan, 2003).

health care; investigated the ways in which race is defined in various federally funded surveys; reviewed the literature on race, prejudice, and discrimination; and examined other literature on survey design, experimental evidence, and statistical analysis.

REPORT ORGANIZATION

This report is divided into three parts. The chapters in Part I provide a conceptual framework for thinking about racial discrimination. Chapter 2 explores the meaning of race as a social construct and provides historical background on the complex issues surrounding race in the United States and how it is measured in the decennial census and other federal data collections. Chapter 3 defines discrimination from a social science perspective and explains why we focus on racial discrimination. Our definition of racial discrimination is informed by legal concepts of discrimination, but it also encompasses behaviors and processes that may not be unlawful or easily measured. Chapter 4 provides a framework for understanding how racial discrimination may operate. As the discussion indicates, there are different ways in which discrimination can occur and various mechanisms that can result in discriminatory behavior. Identifying various sources of discrimination is a crucial first step in developing theories or models of discrimination and using them to guide data collection and research for measuring the presence and extent of different types of discrimination.

The chapters in Part II examine methodological approaches to measuring discrimination and the advantages, limitations, and best techniques associated with each. Chapter 5 provides a general framework for inferring causation and a brief introduction to some of the topics covered in detail in the chapters that follow. Chapter 6 focuses on experimental methods, including field and laboratory experiments. Chapter 7 describes the use of statistical analysis of observational data to measure discrimination, reviewing the necessary assumptions and potential credibility of various approaches. Chapter 8 focuses on approaches employing attitudinal and behavioral indicators of discrimination, including methods based on survey data and administrative records. Each of these chapters describes specific approaches and the situations in which they can be implemented and may be appropriate. Where possible, we also attempt to identify more and less credible approaches, providing guidance for future scholars seeking to use the most effective methods. Chapter 9 at the end of Part II addresses issues of racial profiling, as an illustration of an area in which measuring discrimination is difficult.

The chapters in Part III present the panel's priorities for data collection and research for improved measures of race and racial discrimination. Chapter 10 describes the data collected by federal statistical and administrative agencies that may support analysis of racial discrimination and its effects. The discussion focuses on concepts and measures of race and ethnicity in federal data sources, how different measures may affect distributions and consequent analyses for racial and ethnic groups, and research that is needed to improve federal measures. Chapter 11 considers the nature of cumulative effects of discrimination within and across multiple domains, seeking to identify techniques that can be used to provide a fuller measure of the impact of discrimination when it occurs over time and in more than one social arena. Little empirical work has been done on cumulative discrimination, so research and data collection in this area are important to pursue.

Finally, Chapter 12 suggests next steps for program and research agencies to build a research agenda that is directed to priority needs for measuring racial discrimination. The aim of the chapter is not to develop a detailed agenda per se, which is beyond the panel's scope and resources, but to suggest a series of steps whereby agencies may identify priority research topics; evaluate them for feasibility and cost-effectiveness; and bring to bear the necessary conceptual frameworks, research methods, and data. Whether conducting research from a policy perspective or more basic research, it will be important to support multidisciplinary studies that draw on a range of methodologies and data sources.

The report ends with two appendixes: Appendix A presents the agenda for the Workshop on Measuring Racial Disparities and Discrimination in Elementary and Secondary Education held by the panel in July 2002; Appendix B provides biographical sketches of the panel members and staff.

Part I

Concepts

There is substantial evidence of differential outcomes for different racial and ethnic groups in many social and economic arenas in America today, ranging from hiring and promotion in the labor market, to loan approvals in mortgage lending, to rates of arrest and incarceration in the criminal justice system. These disparities—and others—describe social conditions that most Americans believe deserve some measure of attention. To understand such conditions and fashion appropriate responses, it is important to assess whether and how racial discrimination, along with other factors, may contribute to observed disparities among racial and ethnic groups.

Research in social psychology suggests that categorizing individuals on the basis of salient, observable characteristics such as race, gender, age, and even patterns of dress and speech is inevitable, occurs automatically, and activates biases associated with these characteristics (Allport, 1954; Brewer and Brown, 1998; Devine, 1989, 2001; Fiske, 1998). Some researchers posit that automatic categorization will fail to elicit biased responses among those motivated not to be prejudiced (Devine, 1989, 2001), yet people regularly use such categories to make distinctions and sometimes to perpetuate social inequalities among different groups. Although people categorize others in various ways, the focus in this report is on racial and ethnic distinctions and identifying methods that may make it possible to measure the presence and extent of racial and ethnic discrimination in a social or economic domain.

Before reviewing measurement methods, it is necessary first to define *race* and *racial discrimination* and to outline theories or models of how different types and mechanisms of discrimination may operate in various arenas. A clear conceptual framework is needed to guide appropriate data collection and analysis for measurement and to identify key assumptions of the underlying model. This part of the report is intended to provide that framework.

Chapter 2 briefly reviews biological and social concepts of race and adopts a social–cognitive definition in which race is based on observable physical features and associated characteristics that have acquired socially significant meaning. As such, social categorizations of race evolve over time and differ in different societies and contexts. The chapter presents background on the history and meaning of race in the United States and summarizes the federal government's racial and ethnic categories for data collection. The discussion reviews the ambiguities that complicate the definition and measurement of race.

Chapter 3 provides a social science definition of racial discrimination with two components that correspond to legal concepts of disparate treatment discrimination and disparate impact discrimination. Each component goes beyond a strictly legal definition to include some individual behaviors and organizational processes that may not be unlawful but that have discriminatory effects; that is, they result in adverse consequences for racial groups. We explain why we focus in this report on racial discrimination against disadvantaged groups, and we provide evidence of racial disparities in outcomes across various social and economic domains. Racial disparities can occur for many reasons; their existence motivates our examination of social science methods that can help determine the role that race-based discrimination may play in those differences.

In Chapter 4, we provide theories or models of racial discrimination. To be able to measure racial discrimination of a particular kind, it is necessary to have a theory or model of how such discrimination might occur and what its effects might look like. The theory or model, in turn, specifies the data that are needed to test the theory, appropriate methods for analyzing the data, and the assumptions that the data and analysis must satisfy in order to support a finding of discrimination. Without such a theory, analysts may produce results that are hard to interpret and do not stand up to rigorous scrutiny. The chapter discusses four types of discrimination and the various mechanisms that may lead to such discrimination. It then discusses how these discriminatory behaviors and practices might operate within the domains of education, employment, housing, criminal justice, and health. Finally, the chapter briefly discusses concepts of cumulative discrimination across domains and over time.

2

Defining Race

The focus on measuring racial discrimination in this report raises an initial question of "What is race?" Defining race is a task far more complex than can be accomplished in this chapter. In fact, there is little consensus on what race actually means (Alba, 1992; for discussions on the meaning of race, see Anderson and Fienberg, 2000; Appiah, 1992; Fredrickson, 2002; Jones, 1997; Loury, 2002; Omi, 2001; Winant, 2001). Therefore, we only briefly describe ways in which race (and ethnicity) may be defined, rather than attempting an in-depth analysis.

In this chapter, we first summarize biological and social concepts of race. Next, we present background on the history and meaning of race (and ethnicity) in the United States. We then briefly discuss the federal government's racial and ethnic categories for data collection (which are examined more fully in Chapter 10) and highlight the ambiguities that complicate the definition and measurement of race. We conclude that, for analyzing discrimination and its effects on social, economic, political, and other outcomes for population groups, race is best thought of as a social construct that evolves over time. The discussion here and in the next two chapters makes clear that data on race and ethnicity are necessary—despite measurement problems—for monitoring and analyzing evolving differences and trends among groups in the U.S. population.

BIOLOGICAL DEFINITION

Biological classifications of race were first developed from the work of eighteenth-century naturalists who studied population groups in what had been relatively isolated geographic areas (for further discussion, see Montagu, 1972; Zuberi, 2001). The term "race" was used to distinguish populations in different areas on the basis of differing physical characteristics that had developed over time, such as skin color, facial features, and other characteristics (van den Berghe, 1967; Zuckerman, 1990).

Recently, genetics researchers have found evidence of genetic clusters that correspond to geographically similar populations and yield the kinds of phenotypic variations that have been used to construct concepts of race. Rosenberg et al. (2002) report on a study of 1,056 individuals from 52 different populations. The researchers found that a "soft" classification method using no a priori information on population groups identified six genetic clusters, five of which correspond directly to major geographic regions, as well as subclusters corresponding to specific populations. However, they concluded that within-population differences accounted for 93–95 percent of genetic variation in these individuals, supporting the argument that there are only small genetic differences among geographically different groups.

Although not all scientists agree (see Crow, 2002; Mayr, 2002; van den Berghe, 1967; Zuckerman, 1990), many critics deny that meaningful distinctions among contemporary human groups can be derived from a biological notion of race (see Cavalli-Sforza, 2000; Mead et al., 1968; Omi, 2001). At this point, science has not identified a set of genes that correspond with social conceptions of race. The panel offers no further discussion of any such biological components and focuses on race as a socially constructed concept.

SOCIAL CONSTRUCTION OF RACE

In virtually all human societies, people take note of and assign significance to the physical characteristics of others, such as skin color, hair texture, and distinctive features. Race becomes socially significant when members of a society routinely divide people into groups based on the possession of these characteristics. These characteristics become socially significant when members of a society routinely use them to establish racial categories into which people are classified on the basis of their own or their ancestors' physical characteristics and when, in turn, these categorizations elicit differing social perceptions, attitudes, and behaviors toward each group (see, e.g., Hollinger, 2000; Loury, 2002; Smelser et al., 2001).

The notion that race is about *embodied social signification* may be referred to as the *social–cognitive approach* to thinking about race (Fiske and Taylor, 1991; Loury, 2002). It is important to understand that this approach is conceptually distinct from biological–taxonomic notions of racial classification. No *objective* racial taxonomy need be valid to warrant the *subjective* use of racial classifications. In the social–cognitive sense, "races" may be identified in a society, acknowledged over generations, and believed to be biologically determined even though such groups may not exist in the biological–taxonomic sense.

Recent behavioral and social science evidence supports the social–cognitive notion that race is a construct based on observable physical characteristics (e.g., skin color) that have acquired socially significant meaning (see Banton, 1983; Loury, 2002; Omi and Winant, 1986). In addition to physical features, ascribed and other characteristics such as given name, dress, and diet may also contribute to racial categorizations (see, e.g., Nagel, 1994). Cultural factors, such as language, religion, and nationality, have more often been used to refer to ethnicity—that is, groups of people who share a common cultural heritage, such as various European immigrant groups in the United States (Bobo, 2001).[1]

The social meaning given to racial classifications activates beliefs and assumptions about individuals in a particular racial category. Consequently, if someone is perceived or identifies himself or herself as belonging to the African American or another racial group—regardless of the person's precise physical or other characteristics—that classification creates a social reality that can have real and enduring consequences. For instance, racial classification can affect access to resources (e.g., education, health care, and jobs), the distribution of income and wealth, political power, residential living patterns, and interpersonal relationships. Moreover, the consequences of racial classification over time can create boundaries among racially defined groups that affect people today.

RACE IN THE UNITED STATES

We begin our discussion of race in the United States with its founding in 1789.[2] The founding document, the U.S. Constitution, accorded de facto recognition to white and nonwhite racial categories in order to assign political representation to the states (Anderson and Fienberg, 2000). During the 1787 Constitutional Convention, northern and southern states compromised on counting slaves as three-fifths of a person for purposes of congres-

[1]Nationality is also sometimes used, as in the U.S. census race question for Asians, to distinguish subcategories for a broad racial group. The distinctions between concepts of race and ethnicity are not clear-cut.

[2]For the origins of modern concepts of race and racism in Europe and the influence of European concepts on the North American colonies, see Anderson (1983), Blaut (1993), Frederickson (2002), Graham (1990), Hannaford (1996), Higginbotham (1996), Klein (1999), Northrup (1994), O'Callaghan (1980), and Winant (2001).

sional reapportionment. In addition, Indians "not taxed" were to be excluded from the reapportionment counts. The compromise on the treatment of slaves was key to the establishment of the new government; it was implemented in the U.S. decennial census, first conducted in 1790, when enumerators were instructed to classify people as white, other free person, or slave and to exclude Indians not taxed. Because all slaves were treated as a single racial group (i.e., black), the enumeration of people by their civil status effectively produced a racial classification of whites, American Indians, and blacks (free and enslaved).

Although slavery was abolished in the mid-nineteenth century, for more than 100 years federal and state laws and court decisions (e.g., Jim Crow restrictions put in place in southern states and the 1882 Chinese Exclusion Act) upheld racial classifications as the basis for unequal treatment of groups, serving to maintain dominant and subordinate racial groupings in U.S. society (Feagin and Feagin, 1996).[3] During this period, the concepts of "white" and "nonwhite" were defined in laws and customs to exclude people from white status if they had even a small amount of nonwhite blood. Thus, progeny of black and white unions were invariably classified as "black," regardless of their skin color or appearance. Originating in the South, the so-called one-drop rule was associated with rigid social segregation and economic exploitation.

This narrow concept of whiteness even excluded many immigrant groups that are now classified as white but were not so in the nineteenth and early twentieth centuries. Being white once signified Anglo-Saxon heritage; thus, aside from the British, Dutch, Germans, Scandinavians, and Scotch-Irish, many newly arrived Southern and Eastern Europeans, among others, had to struggle to be defined as "white" in America. Irish immigrants (Ignatiev, 1995), as well as Jews, Italians, and Poles (Roediger, 1991), were often treated negatively because of their ethnic origin. Initially, ethnic and cultural traits set these groups apart from the mainstream as "nonwhites," but over time they acquired many traits of the larger society and were assimilated into "white" U.S. culture. Immutable physical characteristics, however, made it difficult for many non-European groups to follow suit. Thus, such groups as Chinese or Japanese immigrants were denied

[3]Some references on exclusion and discrimination against major nonwhite groups in the United States are, for a range of societal groups, Fiske et al. (2002); for African Americans, Beck and Tolnay (1990), Brundage (1993), Jones (1997), National Research Council (1989), Plous and Williams (1995), Stephan (1985), Stephan and Rosenfield (1982); for American Indians, Blackwell and Mehaffey (1983), Thornton (1987, 2001); for Asian Americans, Hurh and Kim (1989), Ichioka (1977), Kitano and Sue (1973), Sue and Okazaki (1990), Sue et al. (1975); for Hispanic Americans, Camarillo and Bonilla (2001), Huddy and Virtanen (1995).

citizenship for many decades, and Chinese immigration was totally prohibited from 1882 to World War II.[4]

In contrast to U.S. notions of race, Latin American and Caribbean societies have generally lacked a dichotomous classification of people as white or nonwhite; they also take account of social class as well as appearance in determining degrees of "whiteness" (Degler, 1971; Toplin, 1974). Although U.S. society has generally not recognized gradations of whiteness, it has often recognized gradations within broad nonwhite racial categories on the basis of skin color and ancestry. For example, censuses from 1850 through 1890 and again in 1910 and 1920 included one or more black racial subcategories, such as black, mulatto, quadroon, and octoroon. Censuses since 1890 have also distinguished subgroups of Asians on the basis of national origin (see section on "Racial Categories in Federal Statistics" below).

There is social scientific research on skin color differentiation (and discrimination) whereby members of a racial group are further distinguished (and treated differently) on the basis of their skin tone (e.g., light-skinned versus dark-skinned African Americans). Empirical accounts of this type of differentiation within racial categories include Blair et al. (2002), Keith and Herring (1991), Krieger (2000), Maddox and Gray (2002), and Thompson and Keith (2001). However, we focus on the broad classifications, which, in the United States, have carried social meaning and consequences for all their members.

Most recently in the United States, stigmas attached to some nonwhite groups have appeared to diminish, and there has been increased interest by such groups as Native Americans and people of mixed race to identify themselves as such. In addition, increased immigration from Latino and Caribbean countries has highlighted ambiguities in the measurement of race for these populations. Results from recent censuses document that many Hispanic groups think of their ethnicity (as it is termed in the census) as a racial category. These people often check "other race" rather than a specifically named category, or they do not answer the race question at all. We discuss these developments, which underscore the fluid and socially and politically influenced nature of racial classification, in the next section.

RACIAL CATEGORIES IN FEDERAL STATISTICS

The meaning of race in the United States has shaped and been shaped by the data collected and reported by federal agencies in the census, house-

[4]Recent data on segregation and intermarriage suggest that some Asian and Latino groups, after many years, may be achieving a de facto status as "honorary whites" (Charles, 2001; Farley, 1996).

hold surveys, and administrative records systems. These data are used for many purposes, including to describe and analyze social, economic, and other differences among racial and ethnic groups and to monitor compliance with civil rights laws.

Decennial Census

The U.S. census, as noted above, has included questions on race in every census beginning in 1790 (see Chapter 10 for a detailed discussion). Reflecting evolving conceptions of race and the political power of different groups over time, the racial categories in the census have changed from one decade to the next (see Table 2-1). The census has also often included additional items related to racial categorization, such as questions on national origin, birthplace of parents, language spoken in the home, and ancestry.

The identification of Hispanic origin in the census has varied substantially over time. In 1930 the census included "Mexican" as a separate nonwhite racial category; but after protests from the Mexican government, the category was omitted in subsequent censuses. Prior to 1970, there was no nationwide standard for identifying people of Latin American origin. From 1940 to 1970, the census coded people in five southwestern states who reported Spanish language or surname as "of Spanish origin," and in 1960 and 1970 the census tabulated people of Puerto Rican birth or heritage in three northeastern states.

The 1970 census included a Hispanic origin item that was asked of a 5 percent sample of the U.S. population. Respondents were to indicate whether their origin or descent was Mexican, Puerto Rican, Cuban, Central or South American, or some other Spanish origin. Subsequent censuses have asked everyone whether they are of Hispanic origin and to indicate a specific subgroup. The ethnic origin question is separate from the race question; thus, Hispanics can be of any race.

We are in agreement that "Hispanic" is an ethnicity and not a race in the way that Americans have conceptualized such racial groups as whites, blacks, and Asians. However, disadvantaged Hispanics face many of the same barriers to full participation in U.S. society as disadvantaged racial groups. Moreover, from census reporting patterns, significant proportions of Hispanics consider their origin to be synonymous with their race. For these reasons, we include Hispanics when discussing discrimination against different racial groups in this report.

Federal Classification Standards

Since 1977 the U.S. Office of Management and Budget (OMB) has provided federal statistical and program administration agencies with classifi-

TABLE 2-1 Racial Categories in the U.S. Census, 1790–2000

Year	Category
1790	Free Whites, Other Free Persons, and Slaves
1800 and 1810	Free Whites; Other Free Persons, except Indians not taxed; and Slaves
1820	Free Whites, Slaves, Free Colored Persons, and other persons, except Indians not taxed
1830 and 1840	Free White Persons, Slaves, Free Colored Persons
1850	White, Black, and Mulatto
1860	White, Black, Mulatto, and Indian
1870 and 1880	White, Black, Mulatto, Chinese, and Indian
1890	White, Black, Mulatto, Quadroon, Octoroon, Chinese, Japanese, and Indian
1900	White, Black, Chinese, Japanese, and Indian
1910	White, Black, Mulatto, Chinese, Japanese, Indian, Other (plus write-in)
1920	White, Black, Mulatto, Indian, Chinese, Japanese, Filipino, Hindu, Korean, and Other (plus write-in)
1930	White, Negro, Mexican, Indian, Chinese, Japanese, Filipino, Hindu, Korean (Other races, spell out in full)
1940	White, Negro, Indian, Chinese, Japanese, Filipino, Hindu, Korean (Other races, spell out in full)
1950	White, Negro, Indian, Japanese, Chinese, Filipino (Other races, spell out)
1960	White, Negro, American Indian, Japanese, Chinese, Filipino, Hawaiian, Part Hawaiian, Aleut, Eskimo
1970	White, Negro or Black, Indian (American), Japanese, Chinese, Filipino, Hawaiian, Korean, Other (print race)
1980	White, Negro, Japanese, Chinese, Filipino, Korean, Vietnamese, Indian (American), Asian Indian, Hawaiian, Guamanian, Samoan, Eskimo, Aleut, Other (specify)
1990	White, Black, Indian (American), Eskimo, Aleut, Chinese, Filipino, Hawaiian, Korean, Vietnamese, Japanese, Asian Indian, Samoan, Guamanian, Other Asian Pacific Islander, Other race
2000	White; Black, African American, or Negro; American Indian or Alaska Native (specify tribe); Asian Indian; Chinese; Filipino; Other Asian (print race); Japanese; Korean; Vietnamese; Hawaiian; Guamanian or Chamorro; Samoan; Other Pacific Islander (print race); Some other race (individuals who consider themselves multiracial can choose two or more races)

SOURCES: 1790–1990 data adapted from Anderson and Fienberg (2000: Tables 3 and 4) and 2000 data from U.S. Census Bureau (2001a).

cation standards for collecting and reporting data on race and ethnicity (these standards were most recently revised in 1997). OMB currently defines five major racial categories for use in federal data collection: American Indian or Alaska Native; Asian; black or African American; Native Hawaiian and other Pacific Islander; and white. In addition, there are two ethnic categories: Hispanic or Latino and Not Hispanic or Latino. The OMB standards state a preference for separate questions to ascertain race and ethnicity, but agencies may use a combined question in which Hispanic is treated as a race. The standards also state a preference for self-reporting of race and ethnicity as opposed to reporting by an observer.

Under the 1997 standards, respondents for the first time have the option of checking more than one racial category. This option reflects changes in the nation's diversity as a result of immigration and intermarriage among different racial groups. Although a "multiracial" option was considered, many critics believed it would hamper the ability to identify the racial groups to which "multiracial" respondents belonged (Harrison, 2002). Other critics argued that "mark one or more races" might complicate racial classification as well (see Chapter 10 for a history of the OMB categories and recent revision).

OMB emphasizes that its classification standards are designed to monitor adherence to and enforce civil rights laws. "The categories in this classification are social-political constructs and should not be interpreted as being scientific or anthropological in nature" (Office of Information and Regulatory Affairs, 1997:58,788).

Federal statistical and program evaluation and administrative agencies use the OMB standards to count and classify the U.S. population. Administrative data on race are collected by such agencies as the U.S. Department of Education and others to monitor equal access to opportunity in social and economic domains for groups that have experienced discrimination and that may continue to face unequal treatment because of their race or ethnic origin. Census and household survey data on race and ethnicity are collected by the Census Bureau and other statistical agencies and are widely used by federal, state, and local government agencies, private firms, nonprofit organizations, academic researchers, the media, and the general public. Uses of these data include redrawing congressional and state legislative district boundaries; calculating and analyzing vital demographic information (e.g., birth rates, infant mortality rates); monitoring compliance by employers with equal opportunity employment laws; and many other applications for research and program planning, implementation, and evaluation.

The federal classification system is not an attempt to categorize all the ethnic and racial groups in the United States. For instance, the classification system does not support separate identification of many subgroups within racial populations (e.g., Caribbean-born blacks versus U.S.-born African Americans, or foreign-born versus native-born Asians) or of the large numbers of immigrant groups to the United States. Federal agencies such as the Census Bureau can use more detailed subcategories of race (e.g., Vietnamese, Korean, or Filipino for Asian groups) as long as they collapse to the basic five categories cited above. The census is also allowed to use a category of "other race."

MEASUREMENT ISSUES

The Ambiguity of Race

As a social–cognitive construct, the meaning of race in the United States has changed and will likely continue to change over time with changing sociopolitical norms, economic patterns, and waves of immigration (e.g., the assimilation of some European immigrant groups from "nonwhite" to "white" status in the first half of the twentieth century and the growing acknowledgment of mixed-race origins in the twenty-first century). Moreover, race has and may continue to have different meanings for different groups, sometimes overlapping and sometimes not (Lieberman, 1993). For instance, some Hispanics, who can be of any race in the OMB classification system, identify themselves primarily by ethnic or national origin (e.g., Mexican, Cuban, or Puerto Rican) (de la Garza et al., 1992).[5] In contrast, other Hispanics consider Hispanic or Latino to be a race on a par with black, white, Asian, and American Indian (Denton and Massey, 1989; Harris, 2002; U.S. Office of Management and Budget, 1997). Furthermore, whereas historically Americans have most often viewed racial categories as mutually exclusive, Hispanics have tended to see race along a continuum (Nobles, 2000). Thus, there is research evidence that many Latin American and Latin Caribbean immigrants who come to the United States see themselves as being of mixed origin, most commonly European and American Indian or European and African.[6]

Shifts in societal views on race, political pressures from different groups, increasing diversity in the country's population, and consequent changes in data collection standards and practices add ambiguity to the way we understand race and interpret data on race. Two specific measurement problems are inconsistent reporting for individuals and groups, currently and over time, and different data collection practices, such as self-reporting in surveys and, frequently, reporting by others in administrative records systems.

---

[5]Although this identification may vary among Hispanics who are foreign versus native born, first versus second generation, and so on.

[6]Over time and with greater exposure to U.S. culture and society, some Latin American immigrants and their children come to understand the Anglo-American conceptualization of race and shift to the U.S. taxonomy. Indeed, rising socioeconomic status, multiple generations born in the United States, and time spent in the United States reliably predict racial identification as "white," and such identification is often used as an indicator of cultural assimilation (Massey and Denton, 1992).

Some researchers have suggested that multiple indicators are needed to fully understand racial categorization in American society today.

Inconsistent Reporting

Population groups and individuals vary in their consistency of reporting race when comparing surveys across time and with each other. In particular, there has always been considerable confusion regarding the responses of Latin American and Caribbean groups to the separate race question used in the census and surveys. Many Latin Americans and Caribbeans reject the racial categories on the census, select "other race," and write in a word that to them best describes their racial identity.[7] For instance, Latin Caribbeans might use "moreno," "trigeño," or "Boricua," and Central and South Americans might use "mestizo," "la raza," "Mexicano," or some other term denoting a mix of races (Denton and Massey, 1989). From 1940 through 1970, the Census Bureau assumed that such responses were incorrect because Hispanic respondents misunderstood the race question. Accordingly, all such responses were recoded as "white." In 1980, however, the Census Bureau for the first time accepted such responses, grouped them into one category, and treated Hispanic data as "racial" data (i.e., responses were recoded as "Spanish race")—a practice generally continued in 1990 and, with greater complexity, in 2000. This practice further adds to the confusion about how to classify Hispanics as a group.

Different question formats or wording can also affect responses and enumeration of Hispanics. For instance, according to various studies, when questions about race and Hispanic origin are reversed so that the race question follows the ethnic origin question (as was done in the 2000 census), response rates for Hispanic origin are higher, and fewer respondents choose "other race" (Anderson and Fienberg, 2000; Bates et al., 1994; del Pinal, 2003; Martin et al., 1990; Tucker and Kojetin, 1996; see also Harris, 2002). For another example, a 1995 supplement to the Current Population Survey found that a smaller proportion of respondents identified themselves as Hispanic when a combined race and Hispanic origin question was used, compared with having separate questions on race and ethnicity.[8] Finally,

[7]In the 2000 census, 97 percent of people reporting "some other race" were of Hispanic origin, and about one-half of Hispanics either marked "some other race" or marked two or more races, most often a combination of "white" and "some other race" (del Pinal, 2003).

[8]More details on the Tucker and Kojetin (1996) study can be found in Chapter 9. A Race and Ethnic Targeted Test survey conducted by the Census Bureau in 1996 tested a combined race and ethnicity question that, in contrast to the 1995 Current Population Survey supplement, allowed respondents to mark one or more categories. The 1996 survey found no decline in reporting of Hispanic origin in comparison with the two-question format. Nonresponse was also significantly reduced in the combined format (Hirschman et al., 2000).

studies that have compared responses for the same individuals with separate surveys show high rates of consistency for reporting of Hispanic origin but low rates of consistency for reporting of race by Hispanics (see del Pinal, 2003, for a summary of such studies conducted in conjunction with the 2000 census).

Inconsistent reporting of race on surveys is also problematic for other groups, although not to the same extent as for Hispanics. For American Indian and Native Hawaiian or other Pacific Islander groups—both small populations—rates of inconsistent reporting across surveys can be high. For instance, census counts of American Indians and Alaska Natives increased dramatically following the 1950 census—by 51 percent from 1950 to 1960, 50 percent from 1960 to 1970, and 71 percent from 1970 to 1980. Factors driving the increases included not only increased life expectancy but also the change from observer to self-reporting in the census, growing ethnic pride, and reduction in stigma from being identified as anything other than white. There are no data on how many people changed their racial category between censuses, but such people likely accounted for a major part of the growth in the American Indian population after 1950 (Snipp, 2000).

High levels of inconsistent reporting also characterized people who chose more than one race in the 2000 census, which was the first census to allow multiracial identification. Overall, only 2.4 percent of the population marked more than one race, but 8 percent of children ages 0 to 4 had two or more races marked, indicating that the multiracial population is likely to grow in number.[9] However, nearly one-third of multiracial respondents were of Hispanic origin, many of whom checked "white" and "some other race." Also, results from the National Health Interview Survey suggest that, if prompted, a majority of people choosing more than one race will select one primary racial category (see Sondik et al., 2000).

Overall, these results illustrate how respondents have different concepts for race and ethnicity, leading to subjective responses (Anderson and Fienberg, 2000). Subjectivity and ambiguity of responses make it difficult to collect complete and reliably reported information on race and ethnicity. Furthermore, with changes to questions or to category labels for some groups over time (e.g., the Asian or Pacific Islander category was split into two separate categories), differences in who reports the race of an individual (the individual, another household member, or an observer), and changing political reasons for identifying with a particular race (e.g., civil rights enforcement, collective identity), responses can be inconsistent and difficult to interpret.

[9]There was also evidence of a rise in interracial births between 1977 (2.0 percent) and 1998 (5.3 percent), which suggests that the multiracial population is growing (National Center for Health Statistics, 2001).

Despite these problems, it is important to note that 98 percent of the U.S. population identified with one race in the 2000 census (U.S. Census Bureau, 2001a; see also Table 10-2 in Chapter 10). Moreover, studies conducted in conjunction with the census (see del Pinal, 2003) find high levels of consistent reporting for people reporting African American, Asian, or white race, even when the data are collected by different methods (e.g., computer-assisted personal interviewing versus mail response) and use different question formats.

Self-Identification of Race

According to OMB's 1997 revised standards, self-identification or self-reporting is the preferred method of collecting data on race and ethnicity (U.S. Office of Management and Budget, 1997). Self-identification is used in most surveys (and has been used in the census for most of the population since 1960), but many administrative racial identification forms are filled out by someone other than the person being surveyed (e.g., by a parent, survey administrator, health care worker, or police officer).

Some people believe self-identification is the only reasonable method because it allows people to express their own racial identity (see Harris, 2002). Others argue against self-identification because they believe federal racial data, if used to monitor and enforce civil rights, should capture the observer's report of an individual's race—after all, people are most often discriminated against on the basis of observers' beliefs. The revised OMB standards state that, although the data are used for civil rights enforcement, the government should not tell individuals how to classify their race (Harris, 2002; U.S. Office of Management and Budget, 1997).

Using different approaches to identify and report race and ethnicity can make it difficult to compare racial categorizations across time and among data sets. Moreover, reporting procedures are not often the same even within a single data set. For example, 78 percent of occupied households mailed back a questionnaire in the 2000 census, but the remaining forms were obtained by enumerators in follow-up, sometimes from neighbors or landlords (Stackhouse and Brady, 2003). In addition, one person typically fills out a census questionnaire for an entire household, which can result in different racial categorizations from those that each household member would have chosen individually.

Multiple Indicators of Racial Identification

Harris (2002) argues that classifying race is a social process that varies across contexts and observers. He uses a matrix to illustrate the multidimensional, socially constructed nature of racial classification (see Figure

	Genotype/Ancestry	Phenotype	Culture
Internal	1 "I know that my background is X"	2 "I know that I look X"	3 "I know that I feel/act X"
Expressed	4 "My background is X"	5 "I look X"	6 "I feel/act X"
External	7 "He/she has X ancestry"	8 "He/she looks X"	9 "He/she acts X" "He/she thinks he/she is X"

FIGURE 2-1 A matrix of race.
SOURCE: Harris (2002).

2-1). To determine an individual's race, people may use one or more ancestry or biological bases, phenotypic or physical characteristics, and cultural bases, such as ideology and language. Furthermore, racial classifications for an individual may differ according to the perspective of the person making the classification: internal (self-classification based on an individual's beliefs about his or her own race); expressed (self-classification based on how an individual presents his or her race to others—e.g., choosing not to identify as a member of a nonwhite group to avoid stigmatization); and external (classification by observers based on their views of an individual's race). These dimensions are not mutually exclusive; they all interact within a social context. Thus, a mixed-race individual may identify herself as multiracial in private settings but express her dominant race in public and be classified in different categories by different observers.

Obtaining multiple indicators of racial identification would likely provide helpful data to inform racial classification and analysis. As Harris notes, however, indicators for race along these lines are not available in most current data sets. To collect these data, it would be necessary to add specific additional categories and observations, still further complicating the measurement and analysis of race data.

SUMMARY AND CONCLUSION

There is no single concept of race. Rather, race is a complex concept, best viewed for social science purposes as a subjective social construct based on observed or ascribed characteristics that have acquired socially significant meaning. Indeed, for the purpose of measuring racial discrimination, a social–cognitive concept of race is integral to meaningful analysis. The reason is that racial discrimination historically has been and continues today to be a phenomenon of social attitudes and behaviors, stemming from people's perceptions (see Chapters 3 and 4). There is no scientifically objective information that people use or can use as a basis for creating unambiguous, consistent racial classifications that have social meaning and effects.

In the United States, ways in which different populations think about their own and others' racial status have been affected over time by changing patterns of immigration, social and economic change, and changes in societal norms and government policies. The subjectivity of race and the heterogeneity within population groups add further ambiguity to classifying different populations by race. Overall, federal racial categories provide only a partial picture of the heterogeneity and growing diversity of the U.S. population and of the complexity of racial classification. Moreover, factors related to survey and administrative records design and implementation—such as changes in racial categories, methods of reporting data (self-reports or observer reports), and allocation rules for single races and multiple-race combinations—have implications for the collection, use, and interpretation of data on race (especially when attempting to compare data for racial categories in different data sets). Yet most people continue to identify with a single race, and consistency of reporting is high for some major racial groups.

Conclusion: *For the purpose of understanding and measuring racial discrimination, race should be viewed as a social construct that evolves over time. Despite measurement problems, data on race and ethnicity are important to collect (the reasons why are discussed more thoroughly in Chapters 3 and 4).*

3

Defining Discrimination

In the previous chapter we discussed race as a social–cognitive construct that evolves over time and in which racial categories reflect one's own or one's ancestors' physical features and associated characteristics that have acquired socially significant meaning. In this chapter we turn to the concept of racial discrimination, defining it from a social science perspective, which includes not only legal definitions of discrimination but also aspects that go beyond legal concepts. We provide examples of the large and persistent differential outcomes by race in various social and economic domains that make racial discrimination an important topic for social science analysis and motivate our examination of methods for measuring the role that race-based discrimination may play in those differences.

For completeness, we examine the legal definitions of discrimination. Although discrimination is often understood in legal terms because, once it has been identified, legal consequences ensue, our definition encompasses forms of discrimination that may not be explicitly unlawful or easily measured.

A DEFINITION OF RACIAL DISCRIMINATION

In this report, we use a social science definition of racial discrimination that includes two components: (1) *differential treatment on the basis of race* that disadvantages a racial group and (2) *treatment on the basis of inadequately justified factors other than race* that disadvantages a racial group (differential effect). Each component is based on behavior or treatment that disadvantages one racial group over another, yet the two components differ on whether the treatment is based on an individual's race or some other factor that results in a differential racial outcome. As we discuss further below, we are particularly interested in discrimination that disadvantages racial minorities.

The first component of our definition of racial discrimination occurs when a member of one racial group is treated less favorably than a similarly situated member of another racial group and suffers adverse or negative consequences. This definition of discrimination is used in many social science fields (e.g., economics, psychology, sociology) to refer to unequal treatment because of race. Intentional discrimination of this kind is frequently unlawful under either the Constitution or specific legislative prohibitions, such as those in employment, housing, and education. The second component of our definition of racial discrimination includes some instances in which treatment based on inadequately justified factors[1] other than race results in adverse racial consequences, such as a promotion practice that generates differential racial effects. A process with adverse racial consequences may or may not be considered discrimination under the law, depending on whether there is a sufficiently compelling reason for its use and whether there are alternative processes that would not produce racial disparities.[2] In the areas in which this type of discrimination is unlawful, the reason is to curtail the use of unintentional practices that can harm racial minorities, as well as to sanction intentional discrimination that might not be identified because of the difficulty in establishing intent in the legal setting.[3]

The two components of our definition—differential treatment and differential effect discrimination—are related to, but broader than, the standards applied in a large body of case law—*disparate treatment* and *disparate impact discrimination* (see the detailed discussion below in this chapter).[4] Legally defined, disparate treatment racial discrimination occurs when an

---

[1]Inadequately justified factors refer to those factors within a particular domain that are not justified (germane) for the purpose for which they are used.

[2]Because the Constitution does not itself prohibit disparate impact discrimination, governmental actions will be scrutinized only under this second legal theory of discrimination if they are covered by a specific legislative command (see discussion in "The Legal Definition of Discrimination" below).

[3]For example, in *Griggs v. Duke Power Co.* (401 U.S. 424 [1971]), the Supreme Court held that Duke Power Company used high school graduation and standardized testing requirements to mask their policy of giving job preferences to whites and not to blacks (i.e., disparate treatment discrimination). Neither requirement was intended to measure an employee's ability or performance in a particular job or job category within the company.

[4]For clarity, when referring to legal definitions of racial discrimination, we use the terms "disparate treatment" and "disparate impact." References to "discrimination" refer to our two-part definition.

individual is treated less favorably—for example, is not hired for a job—because of his or her race. Disparate impact racial discrimination occurs if a behavior or practice that does not involve race directly has an adverse impact on members of a disadvantaged racial group without a sufficiently compelling reason. An example is an employment practice or policy against hiring job applicants with a criminal arrest record when such a policy results in proportionately fewer hires for disadvantaged racial groups while not significantly advancing any legitimate employer interests. These kinds of practices and policies—whether intentionally or unintentionally harmful—are deemed unlawful unless a sufficiently compelling business reason can be supplied to justify them.

Although our definition encompasses the legal definitions of discrimination, we do not believe that a social science research agenda for measuring discrimination should be limited by those legal definitions. Although many of the issues that we discuss may be relevant to certain debates within the courts, our primary intention in this report is to provide guidance to social science researchers interested in measuring racial discrimination. Therefore, in our definition we allow both categories to include a range of behaviors and processes that are either not explicitly unlawful or not effectively prohibited because of difficulties in measurement or proof (see Chapter 4). For example, subtle forms of discrimination might not be susceptible to legal challenge but fall within the scope of our definition. An example of a subtle form of discrimination (perhaps unintentional) would be when interviewers of job applicants more frequently adopt behaviors (e.g., interrupting, asking fewer questions, using a hectoring tone) that result in poor communication and consequently poorer performance by disadvantaged minority applicants as compared with other applicants. Compared with overt discrimination, it is often more difficult to find proof that subtle discrimination has occurred and to address it legally, even if in theory such subtle discrimination constitutes actionable disparate treatment discrimination.

In addition, many legislative and administrative actions that have a discriminatory impact are not legally prohibited because the constitutional mandate against racial discrimination does not recognize the disparate impact theory of discrimination. Social scientists, however, will still want to ascertain the possibly discriminatory effects of such legally permissible governmental actions. A final example of discrimination's impact that we want to measure as social scientists, but which may not be unlawful, occurs when discriminatory effects cumulate across domains. Discrimination by real estate agents may result in housing segregation, which in turn affects educational quality (because of local tax financing of the schools) and long-term educational and labor market outcomes. Although discriminating real estate agents can be found liable for housing market discrimination, there is no legal mechanism to allocate blame for educational or labor market differences that such discrimination might induce. Yet, as social scientists we want to identify and measure these cross-domain effects.

LIMITING THE DISCUSSION

The experience of discrimination and its consequences may vary with several factors, including the domain in which it occurs (e.g., the labor market, the health care system, the criminal justice system, the housing market); the actors involved (e.g., employers, insurance companies, police officers, mortgage lenders, neighbors); and the targets (e.g., African Americans, whites, Hispanics, American Indians, Asians). Within the scope of our broad definition of discrimination, we focus our analysis on specific aspects of racial discrimination in the United States. We are concerned primarily with discrimination that has adverse social and economic consequences for disadvantaged racial groups. We use the term *disadvantaged racial groups* interchangeably with *minority groups* and *nonwhite groups* and refer to non-Hispanic whites as the majority group. These terms describe the social stratification (rather than the numerical proportions) of different racial groups in the United States. We recognize that racial groups in different communities, institutions, and even countries (e.g., South Africa) can be in the numerical majority but still experience discrimination.

We acknowledge that non-Hispanic whites may face discrimination that results in adverse consequences (so-called reverse discrimination). However, members of disadvantaged groups have more often been discriminated against in various social and economic arenas (Council of Economic Advisors, 1998; National Research Council, 2001a), and ongoing discriminatory practices and policies can undermine efforts to overcome these disadvantages. Therefore, while we do not rule out the possibility of so-called reverse discrimination, we do not address discrimination against non-Hispanic whites in this report.

We refer more often to evidence of racial discrimination by whites against blacks, although we recognize that other racial groups, including whites, as well as some ethnic groups, face discrimination.[5] Primarily, this is a result of the larger literatures on black–white disparities and research to

[5]For example, much of the social psychological literature (e.g., Greenwald et al., 1998; Rudman et al., 1999) shows evidence of implicit prejudices based on categories other than race, such as religious ethnicity (Jewish versus Christian), age (young versus old), and nationality (American versus Soviet or Japanese versus Korean). After September 11, 2001, an increased number of Arab and Middle Eastern men and women reported experiencing discriminatory behavior at airports around the nation.

measure discrimination. In many data sets, sample sizes are too small for analysis of some groups (such as American Indians), or there is no separate identification of groups, such as Asian Americans or Native Hawaiians and other Pacific Islanders, or subgroups, such as Mexican Americans or Puerto Ricans.

Given limited time and space, we primarily discuss racial discrimination in general terms and do not discuss the differences in experiences of discrimination among racial groups, although we recognize that each group has a different historical experience. Furthermore, the broad categories used in most of the data reported here—such as African American or Hispanic—are very heterogeneous in terms of nativity, phenotype, culture, religion, and socioeconomic background. Although important to consider, nuanced attention to these differences is beyond the scope of this report.

We do not address policy issues regarding racial discrimination. For example, we do not discuss the implications or effectiveness or costs of policies intended to alleviate discrimination (e.g., affirmative action or diversity policies). Our charge is to assess social science research methods for measuring racial discrimination. One use of such methods is to assist in policy formulation and evaluation, but discussion of policies as such goes beyond our charge.

One aspect of differential behavior largely beyond the scope of this report is differences in associational choices made by members of different racial groups, such as whom one lives with and marries, whom one's friends might be, and even whom one sits next to at lunch. Issues of associational choice do not fall into our definition of discrimination, although they may have large and adverse effects on differential racial outcomes. Most (though certainly not all) antidiscrimination efforts are focused on those arenas in which there are contracts or explicit markets for the exchange of goods and services. Ideally, equal access to those markets (be they in employment or in housing) would be available to all racial groups. There is neither a legal nor a social tradition of intervening in associational choices as long as those choices are based entirely on individual preferences and not on group-imposed exclusionary policies or practices. It is not always clear when an associational decision is freely chosen and when it is subject to such tight constraints that it might be considered discriminatory. Although important to the broad understanding of racial group differences in our society, these are issues that necessarily lie beyond the mandate of this panel and that we cannot adequately treat in this report.

Finally, our definition of discrimination is based on behaviors and practices, and as such it differs from a definition that also includes prejudiced attitudes and stereotypical beliefs. Discriminatory behaviors and practices may arise from prejudice and stereotyping, but prejudice need not result in differential treatment or differential effect. Similarly, whereas discriminatory behavior in many domains is unlawful, prejudiced attitudes and stereotypical beliefs are not.

DIFFERENTIAL OUTCOMES BY RACE

Evidence of large and persistent differentials in social, economic, and political outcomes among racial and ethnic groups in the United States characterizes virtually every social domain. Indeed, were there not such marked differences, there would be little reason to convene a panel of social scientists to study methods for measuring race-based discrimination. Even though prejudices and stereotyping might be present and individual cases of discrimination might occur, an absence of observable differences in outcomes among racial groups would almost preclude social science measurement of the role of racial discrimination in American society.

To motivate our report, we provide examples of differential outcomes among racial groups in five domains: education, the labor market, the criminal justice system, the housing market and mortgage lending, and health care.[6] In these examples, we draw no conclusions about whether or to what extent differential outcomes by race are caused by discrimination. The magnitude of the differentials in these—and other—domains, however, is a primary reason to be concerned about our ability to identify and measure racial discrimination. Also, the greater the extent to which differential outcomes are the result of discriminatory behaviors or processes, the greater is the likelihood that antidiscriminatory efforts would be needed to reduce these differences.

Education

Racial classification and many factors that are correlated with race (e.g., family structure, parental education, poverty, access to computers, and linguistic diversity) are associated with different educational experiences and levels of educational attainment (Choy, 2002; Lloyd et al., 2002; Mare, 1995). Research shows that blacks, Hispanics, American Indians, and Native Hawaiians and other Pacific Islanders—compared with whites and Asians—are more likely to attend lower-quality schools with fewer teachers and material resources and greater concentrations of poor, homeless, limited English-speaking, and immigrant students (Kahlenberg, 2001; Lee et al., 2001; Natriello et al., 1990; Van Hook, 2002). They are also more likely to have lower test scores, drop out of high school, not graduate from college, and attend lower-ranked programs in higher education (see Na-

[6]We do not look extensively at trends over time in differential outcomes for these domains, which are reviewed elsewhere (e.g., National Research Council, 2001a).

tional Research Council, 2002a). For example, the U.S. Department of Education (2001a) reports that African American and Hispanic students are less likely to have completed advanced levels of math and science coursework compared with Asian and Pacific Islander and white students. However, overall educational attainment may vary substantially among Asian groups—for example, Japanese, Koreans, and Asian Indians versus Cambodians, Laotians, and Hmongs (U.S. Census Bureau, 1993).

Hispanics continue to face obstacles to educational achievement. Between the late 1970s and 1998, they had significantly lower educational attainment and higher dropout rates than both blacks and whites (Hauser et al., 2002). In 2000, 57 percent of Hispanics aged 25 and over had obtained at least a high school degree, compared with 79 percent of blacks and 85 percent of whites (U.S. Census Bureau, 2002). One factor influencing the education gap between whites and Hispanics is the increasing numbers of disadvantaged Hispanic immigrant groups entering the United States. Poor educational outcomes for many Hispanic groups may lead to subsequent disadvantages in social and economic opportunities (e.g., lifetime earnings or civic participation; see Blank, 2001).

Employment and Income

Black Americans are more likely to experience unemployment as teens and adults, to work at lower wages, to have lower wage growth over time, and to accumulate less wealth relative to whites (Altonji and Blank, 1999). Indeed, unemployment rates for blacks are generally twice those for whites. In 2002 the average annual unemployment rate for black workers aged 16 and over (10.3 percent) was nearly twice the overall unemployment rate (5.8 percent) and just over twice the rate for whites (5.1 percent). The unemployment rate for Hispanics was 7.6 percent that year (Bureau of Labor Statistics, 2003).

Median weekly earnings for blacks ($499 in 2002) and Hispanics ($424) are much lower than for whites ($627). The gap between Hispanics and whites has grown at a particularly rapid rate, a fact that may be attributable to differences in educational achievement (Bureau of Labor Statistics, 2003). These earnings differentials are reinforced by substantial differences in the occupational categories in which various racial groups are clustered, with disadvantaged racial groups generally having lower-status as well as lower-wage occupations. Empirical research on labor market outcomes for Asians and American Indians is more limited, reflecting the lack of data on these groups (Altonji and Blank, 1999).[7]

[7]Darity et al. (2001) are an exception to this; they use decennial census data to look at more disaggregated groups.

Perhaps the largest racial differences are observed with respect to wealth, which reflects not just current earnings but cumulative lifetime (and even cross-generational) differences. The average net worth of blacks is just a fifth that of whites (Conley, 1999; Oliver and Shapiro, 1995).

Criminal Justice

Disadvantaged racial groups (particularly blacks) are disproportionately represented in the criminal justice system compared with non-Hispanic whites. Racial differences are largest in the corrections system, in which the incarceration rate for blacks is about eight times that for whites (Blumstein, 1982, 1993). In large part, this differential reflects more frequent arrests of blacks for serious crimes (e.g., murder and robbery), for which the ratio of black to white arrest rates is about 7, relative to less serious crimes (e.g., burglary and drugs), for which the ratio is closer to 3.[8]

In some cases, the punishment for crimes committed by blacks is significantly different from that for similar crimes committed by whites. One reason is lower thresholds for mandatory minimum sanctions for crimes that are more likely to be committed by blacks. This difference is particularly striking in a provision of the federal Anti-Drug Abuse Act of 1986: A mandatory minimum sentence of 5 years is imposed for possession of as little as 5 grams of crack cocaine; in contrast, a possessor of powder cocaine must have at least 500 grams to receive a mandatory minimum sentence of 5 years. In 2000, 85 percent of sentenced crack cocaine offenders, who were sentenced for possessing very small amounts of cocaine, were black, but only 31 percent of sentenced powder cocaine offenders, who had to have large amounts of cocaine to be sentenced, were black (51 percent were Hispanic and 18 percent white). Although the disparate sentencing thresholds are associated with the crime rather than the race of the offender, they have a marked differential racial impact.

Blacks are disproportionately represented not only as offenders but also as victims of crime (Sampson and Lauritsen, 1997; Walker et al., 1996; Weich and Angulo, 2002). Sampson and Lauritsen report that crime victimization rates vary systematically across racial and ethnic groups. Compared with whites, blacks were six times more likely to be murdered in 2000 (see U.S. Department of Justice, 2001).

In criminal justice research, there is a lack of consistent data on crime

---

[8]These ratios are based on arrest data from Table 43 of the FBI's Uniform Crime Reports for 2000 (U.S. Department of Justice, 2000) and 2000 population data from Table 10-2 in Chapter 10 of the present report (with those of "Other Race," who are predominantly Hispanics, being counted as "white" because the arrest reports do not have a separate count for Hispanics).

for Asian Americans and American Indians. However, Weich and Angulo (2002) point out that Asian American youths are far more likely than whites to be transferred to adult courts, convicted in adult courts, and incarcerated in youth and adult prisons. Also, although African Americans are overrepresented in federal and state prisons relative to their proportion in the population (Walker et al., 1996), American Indians actually have the highest incarceration rate for any race: In 1997, 1,083 of every 100,000 American Indians in the United States were incarcerated (Smelser and Baltes, 2001). Moreover, American Indian youths, who are subject to federal rather than state prosecution, often end up facing harsher sentences than if they were subject to state prosecution (Weich and Angulo, 2002). As a result, approximately 60 percent of youths in federal custody are American Indian.

Housing Markets and Mortgage Lending

Housing segregation among black Americans is far greater than among any other identifiable group. For example, blacks are much more likely to live in segregated neighborhoods, to rent rather than own a home, and to have a lower-valued home when they are homeowners (Charles and Hurst, 2002; Massey, 2001). Although legal segregation and exclusion ended in 1968 with the Fair Housing Act, racial disparities in certain neighborhoods and housing markets continue. In addition, disparities in aggregate lending to black and white neighborhoods continue to exist in many communities (for a review of the evidence, see Ladd, 1998; Munnell et al., 1992, 1996; Turner and Skidmore, 1999; Turner et al., 2002a; Wyly and Holloway, 1999). For example, Wyly and Holloway found that applicants were more likely to have their loans approved in Atlanta neighborhoods in which their race was predominant (i.e., blacks were approved more in black neighborhoods and whites in white neighborhoods). Other studies have shown significant differences in the probability of mortgage loan approval by race in Boston (Carr and Megbolugbe, 1993; Munnell et al., 1992) and Milwaukee (Squires and O'Connor, 2001).

Health Care and Health Outcomes

African Americans, Hispanics, Asian Americans, and American Indians and Alaska Natives face large barriers to health care services as compared with whites (for recent reviews, see Institute of Medicine, 2003; Mayberry et al., 2000). These groups tend to experience lower levels of access to care and to receive lower-quality health care (Institute of Medicine, 2003). For instance, African Americans and Hispanics compared with non-Hispanic whites are less likely to receive kidney dialysis or transplants (Epstein et al., 2000), are less likely to receive appropriate cancer diagnostic tests or treatments (Imperato et al., 1996; McMahon et al., 1999), and are more likely to receive less-than-desirable procedures, such as limb amputation for diabetics (Chin et al., 1998; for additional references, see Institute of Medicine, 2003).

Disadvantaged racial groups are also more likely than whites to suffer from adverse health status and outcomes (Institute of Medicine, 2003; Keppel et al., 2002; National Research Council, 2001a). Thus, substantial racial differentials exist for rates of infant mortality, certain cancers, cardiovascular disease, and kidney disease (Keppel et al., 2002). For example, American Indians are more likely than other racial groups to die from diabetes, liver disease and cirrhosis, and unintentional injuries (Institute of Medicine, 2003). There is also considerable evidence that African Americans have disproportionately high levels of hypertension compared with other racial groups (see Anderson, 1989).

Interpreting Differential Outcomes

Differences in outcomes by race do not themselves provide direct evidence for the magnitude or even the presence of racial discrimination in any particular domain. These outcome differences are the result of any number of factors that may or may not include racial discrimination in that domain. For instance, racial disparities in the labor market (e.g., in hiring or wages) may reflect differences in school quality and achievement rather than any racial animus within the labor market per se. (We discuss these issues further in Chapter 11.)

Although racial disparities continue to exist in many domains, both social and legal changes have improved opportunities for many nonwhites in the United States. Recently, the Brookings Institution (2000) reported the federal government's 50 most important achievements in the past 50 years, including expanding the right to vote (ranked 2), promoting equal access to public accommodations (3), reducing workplace discrimination (5), increasing access to postsecondary education (19), and increasing low-income families' access to health care (34). Examples of legislative acts designed to promote equal opportunity and reduce discrimination include the Civil Rights Act of 1964 banning discrimination in employment and in public accommodations; the Voting Rights Act of 1965 (and its subsequent extensions and amendments), allowing full political participation of nonwhite groups once excluded from voting; Federal Executive Order 11246, requiring compliance by government contractors with federal antidiscrimination policies and the development of administrative systems to monitor compliance; and the Fair Housing Act of 1968, banning discrimination in housing.

Nonetheless, differential outcomes by race persist and motivate analysis to understand contributing factors, including the possible role of racial discrimination. Black–white gaps in income, employment, higher education, test scores, housing segregation, health care, and treatment within the criminal justice system are large. Such sizable and persistent differences in outcomes, by themselves, are problematic and important to address. Even if differential outcomes do not in and of themselves prove that discrimination is occurring, they tell us where to look when seeking to assess whether discriminatory behavior occurs in various social arenas. In the example cited above, for instance, racial disparities in the labor market may reflect not only discrimination in that domain at that time (e.g., wage differentials) but also discrimination in earlier interactions (e.g., labor market experience) and in other domains (e.g., education).

Differential outcomes might be less informative if we believed that the groups involved were innately different. Yet, as noted in Chapter 2, scientists have not determined a genetic basis for the socially based racial and ethnic categories in American society—categories whose meaning has changed over time (e.g., the assimilation of previously "nonwhite" European immigrant groups into the "white" category). We can then infer that these differential outcomes reflect deep differences in the historical and current experiences and environment of disadvantaged racial groups versus non-Hispanic whites. For instance, surveys show that nonwhites perceive much greater discrimination toward nonwhite racial groups and experience much more discrimination themselves compared with whites (Bobo, 2001; Morin, 2001; Schuman et al., 1997). Cumulative disadvantage across generations—in access to nutritious food, decent housing, remunerative employment, and secure and stress-free environments—is a possible way to interpret the differences in current outcomes among nonwhite Americans (see discussion in Chapter 11).

THE LEGAL DEFINITION OF DISCRIMINATION

Thus far we have presented a definition of discrimination and examined racial disparities across several domains. As a point of comparison, in this section we look at the legal definitions of discrimination and identify the circumstances under which a legal finding of discriminatory behavior can be made. The law represents an important venue in which racial discrimination is often identified and measured. In a legal setting, once an act has been labeled as discriminatory, legal remedies, both monetary and injunctive, may be awarded.

An elaborate array of federal and state constitutional, statutory, and administrative provisions broadly prohibit discrimination on the basis of race in a vast range of public and private behaviors. A large body of law has developed to give content to this broad prohibition by defining specifically what constitutes impermissible discrimination. Because the foundations of these laws emanate from different jurisdictions and legal authorities, there is no single definition of impermissible racial discrimination; standards depend on the particular jurisdiction or actor involved. Nonetheless, as noted above, two important doctrinal concepts—disparate treatment and disparate impact discrimination—are useful in defining the nature of the legal prohibition. Each is discussed in turn below.

Disparate Treatment Discrimination

The core concept of disparate treatment discrimination emanates from the constitutional requirement of equal protection under the law and is codified in the main federal statute prohibiting racial discrimination in employment—Title VII of the 1964 Civil Rights Act. This statute prohibits an employment practice that affects an individual's employment "because of such individual's race. . . ." Thus, an employer who refuses to hire, fails to promote, or discharges a worker because of his or her race is guilty of disparate treatment discrimination. So, too, is an employer who decides to pay nonwhite workers less than white workers or to discipline the former more heavily for identical conduct.

The language "because of" is interpreted as requiring proof that race was a motivating factor for the employment practice. In theory, the requirement that the discrimination be intentional before it runs afoul of the law may protect an employer who acts without conscious awareness of having discriminated, a phenomenon that the research literature in psychology indicates is common. For this reason, some legal scholars have suggested that the legal theory of intentional discrimination is flawed and should be expanded to prohibit unconscious or negligent acts of discrimination (Allen, 1995; Oppenheimer, 1993).

In practice, however, a defense that the discrimination is "unconscious" is virtually never encountered in employment discrimination litigation, which typically focuses on two issues: (1) the plaintiff's threshold demonstration of racial disparity in treatment and (2) the credibility of the nondiscriminatory reasons for this disparate treatment offered by the employer. Therefore, although in theory any nondiscriminatory reason will constitute a defense against a charge of disparate treatment discrimination, in practice an employer will be more likely to lose the case if the reason does not appear to be sufficiently linked to the plaintiff's lack of ability to perform the job or demonstrated misconduct. Nonetheless, the courts have held that the burden of persuading the court that the employment decision was discriminatory remains with the plaintiff. Even if the plaintiff establishes that the employer's proffered reason for the employment action is not truthful, the employer will prevail if the plaintiff cannot persuade the court that race was a motivating factor.

Once it has been established that an employer has intentionally discriminated on the basis of race, the reason for the differential treatment will ordinarily not be relevant (unless it is pursuant to the implementation of a valid affirmative action plan). Accordingly, intentional racial discrimination will be deemed unlawful whether the employer acted because he or she dislikes nonwhites (say, blacks), prefers a nonblack ethnic group that is consequently favored, or believes that blacks will be on average less productive. Similarly, an employer cannot engage in disparate treatment on the grounds that customers or other employees demand such racial exclusion or would otherwise prefer it.

On the other hand, a decision to locate a plant in a suburb or in a state with a low black (or other nonwhite) or Hispanic population may have serious adverse consequences for potential black (or other nonwhite) and Hispanic employees. But this locational decision, even if motivated by racial animus, will not be prohibited unless it is deemed an "employment practice." Cases focusing on infrequent institutional behaviors (as opposed to regularly implemented procedures) are essentially unknown for both doctrinal and practical reasons, a fact that underscores how potentially significant choices that may be affected by discriminatory motives can impair the employment prospects of minority groups without generating any legal response.

The task of measuring racial discrimination in a legal case often begins with the documentation of various racial disparities in such areas as income, wealth, educational attainment, incarceration or involvement in the criminal justice system, and health. Of course, as noted above, the mere presence of large disparities in some of these measures does not necessarily mean that discrimination exists. For example, in the United States it is well documented that women live far longer than men, but it is rarely thought that discrimination against men explains their substantially higher rates of death. Similarly, men commit suicide and are incarcerated at vastly higher rates than women, yet again discrimination against men is unlikely to play a large explanatory role in these male–female disparities. Moreover, as discussed above, even when a racial or other group disparity is the product of discrimination, it is not necessarily the result of discrimination occurring at the point in time at which the disparity becomes manifest. For example, employers are generally not held liable under Title VII of the 1964 Civil Rights Act for disparities resulting from pre–labor market discrimination against blacks.

Disparate Impact Discrimination

Although disparate treatment was the original conception of unlawful discrimination, in 1971 the Supreme Court established a second, potentially broader notion of discrimination—the disparate impact standard. Under this doctrine, which, like disparate treatment, was judicially crafted in the arena of employment discrimination, the court first asks if an employment practice, even though facially neutral, has an adverse impact on members of a protected group. Once a finding of disparate impact has been made, the court will rule the challenged practice unlawful unless a sufficiently compelling business justification can be supplied for retaining it. The precise legal standard for this justification defense, first legislatively articulated in the Civil Rights Act of 1991, is that the defendant must prove "that the challenged practice is job related for the position in question and consistent with business necessity." Moreover, this justification will be dismissed if the employer's proffered legitimate business interest could be satisfied by another equally effective employment practice having a less racially adverse impact.

Everything from minimum educational requirements to rules against hiring those with arrest records to grooming standards that prohibit beards to certain types of seniority systems has been deemed under certain circumstances to constitute disparate impact discrimination against disadvantaged racial groups in employment. The goal of the doctrine has been to remove artificial barriers that prevent the economic progress of members of protected groups. At the same time, the rationalization of employment processes that has followed in the wake of the development of the disparate impact doctrine may have brought greater fairness to the process of selection of all employees.

Discrimination Law Regarding Governmental Actions

Government actions fall under a somewhat different set of legal rules. The constitutional prohibition against violations of equal protection (directly applied to the states under the Fourteenth Amendment and indirectly applied to the federal government under the due process clause of the Fifth Amendment) prohibits racial classifications unless justified by a "compelling interest" and unless the policy is "narrowly tailored" to serve that interest. However, the definition of racial discrimination for purposes of evaluating constitutional violations is narrower than the two-part legislative standard that governs employment discrimination law.

Specifically, although the equal protection clause prohibits disparate treatment discrimination that fails to have the most compelling societal justification, the Constitution prohibits only intentional discrimination; evidence of disparate impact alone will not establish a violation. Thus, the Constitution does not restrict a government from engaging in acts that harm disadvantaged racial groups unless the harm is caused intentionally. Moreover, knowing that a certain practice will cause harm is not enough to render it an intentional act of discrimination barred by the equal protection clause. As the court has emphasized, a government is not prohibited from acting in spite of harm to members of disadvantaged racial groups; it is banned only from causing harm because of race.

This constitutional interpretation reflects the fact that many neutral governmental actions have predictable effects that either benefit or harm certain racial groups and that allowing all these actions to be challenged on equal protection grounds would make the federal courts the arbiters of a vast array of legislative and executive conduct. For example, the mortgage interest deduction for residential housing disproportionately benefits whites because of their greater housing wealth and possibly dampens investments in other types of productive capital that might generate more jobs that could disproportionately advantage blacks. Similarly, the war on drugs is designed to identify and punish the tens or even hundreds of thousands of workers in the illegal drug trade, a disproportionate number of whom will inevitably be drawn from disadvantaged groups having less abundant opportunities in the legitimate economy. Yet no doctrine of law would permit either of these ostensibly neutral governmental programs to be challenged as racially discriminatory. Similarly, governmental social programs that disproportionately benefit a racial or ethnic group cannot be challenged on that basis alone on equal protection grounds.

Any form of racially preferential treatment by a government entity—whether in giving preference to minority contractors or to minority applicants to state universities—is subject to strict judicial scrutiny. A racially based treatment implemented by government, even if motivated by the desire to promote affirmative action, will violate the Constitution unless it is "narrowly tailored" to serve a "compelling government interest" (*Adarand Constructors, Inc. v. Peña*, 515 U.S. 220, 237-8 [1995]).[9]

---

[9]Governmental actors are constitutionally constrained not to engage in intentional disparate treatment on the basis of race unless the action can withstand strict judicial scrutiny. Specifically, in *Adarand Constructors, Inc. v. Peña*, 515 U.S. 200 (1995), the Supreme Court announced that all racial classifications by government—whether federal, state, or local—are subject to strict judicial scrutiny under the Constitution and can be sustained only if they are "narrowly tailored" to serve a "compelling government interest." The Court "held that, under the equal protection component of the Fifth Amendment's due process clause or under the equal protection clause of the Constitution's Fourteenth Amendment, all racial classifications, imposed by whatever federal, state, or local governmental actor, must be analyzed by a reviewing court under strict scrutiny, that is, such classifications are constitutional only if they are narrowly tailored measures that further compelling governmental interests" (515 U.S. 200). The Supreme Court has recently reaffirmed this holding in its two cases dealing with affirmative action at the University of Michigan [*Grutter v. Bollinger*, 123 S. Ct 2325 (2003); *Gratz v. Bollinger*, 123 S. Ct 2411 (2003)].

SUMMARY

We adopt a broad definition of racial discrimination for use in social science research, which includes individual behaviors and institutional processes but not attitudes or beliefs as such. Our definition includes two components that are related to (but broader than) a large body of case law: differential treatment on the basis of race that disadvantages a racial group and treatment on the basis of inadequately justified factors other than race that disadvantages a racial group (differential effect). In defining discrimination for this report, we focus primarily on discrimination that has harmful consequences for disadvantaged racial minorities.

Our definition is not limited to those actions defined as discriminatory within a legal framework but also encompasses subtle behaviors and processes and cumulative discriminatory effects that may not be explicitly unlawful or easily measured. In the next chapter, we discuss in more detail the possible ways in which discrimination may manifest itself and return to a discussion of when these discriminatory behaviors may or may not be explicitly unlawful.

There is a history of racial exclusion in the United States and a persistence of large disparate outcomes for racial groups across many societal domains. Although such disparities may not in themselves signal the presence of discrimination in any particular domain or event, they are problematic and motivate our work to assess social science analytical methods for measuring the role of racial discrimination in American society today.

4

Theories of Discrimination

In Chapter 3, we developed a two-part definition of racial discrimination: differential treatment on the basis of race that disadvantages a racial group and treatment on the basis of inadequately justified factors other than race that disadvantages a racial group (differential effect). We focus our discussion on discrimination against disadvantaged racial minorities. Our definition encompasses both individual behaviors and institutional practices.

To be able to measure the existence and extent of racial discrimination of a particular kind in a particular social or economic domain, it is necessary to have a theory (or concept or model) of how such discrimination might occur and what its effects might be. The theory or model, in turn, specifies the data that are needed to test the theory, appropriate methods for analyzing the data, and the assumptions that the data and analysis must satisfy in order to support a finding of discrimination. Without such a theory, analysts may conduct studies that do not have interpretable results and do not stand up to rigorous scrutiny.

The purpose of this chapter is to help researchers think through appropriate models of discrimination to guide their choice of data and analytic methods for measurement. We begin by discussing four types of discrimination and the various mechanisms that may lead to such discrimination. The first three types involve behaviors of individuals and organizations: intentional discrimination, subtle discrimination, and statistical profiling. The fourth type involves discriminatory practices embedded in an organizational culture. Next, we compare these discriminatory behaviors and institutional practices with existing legal standards defining discrimination in the courts (as delineated in Chapter 3). We then discuss how these discriminatory behaviors and practices might operate within the domains of education, employment, housing, criminal justice, and health. Finally, we discuss concepts of how cumulative discrimination might operate across domains and over time to produce lasting consequences for disadvantaged racial groups. This chapter is not concerned with identifying the relative importance of the various types of discrimination; rather, it is designed to present a set of conceptual possibilities that can motivate and shape appropriate research study designs.

TYPES OF DISCRIMINATION

Most people's concept of racial discrimination involves explicit, direct hostility expressed by whites toward members of a disadvantaged racial group. Yet discrimination can include more than just direct behavior (such as the denial of employment or rental opportunities); it can also be subtle and unconscious (such as nonverbal hostility in posture or tone of voice). Furthermore, discrimination against an individual may be based on overall assumptions about members of a disadvantaged racial group that are assumed to apply to that individual (i.e., statistical discrimination or profiling). Discrimination may also occur as the result of institutional procedures rather than individual behaviors.

Intentional, Explicit Discrimination

In 1954, Gordon Allport, an early leader in comprehensive social science analysis of prejudice and discrimination, articulated the sequential steps by which an individual behaves negatively toward members of another racial group: verbal antagonism, avoidance, segregation, physical attack, and extermination (Allport, 1954). Each step enables the next, as people learn by doing. In most cases, people do not get to the later steps without receiving support for their behavior in the earlier ones. In this section, we describe these forms of explicit prejudice.

Verbal antagonism includes casual racial slurs and disparaging racial comments, either in or out of the target's presence. By themselves such comments may not be regarded as serious enough to be unlawful (balanced against concerns about freedom of speech), but they constitute a clear form of hostility. Together with nonverbal expressions of antagonism, they can create a hostile environment in schools, workplaces, and neighborhoods (Essed, 1997; Feagin, 1991).

Verbal and nonverbal hostility are first steps on a continuum of interracial harm-doing. In laboratory experiments (see Chapter 6 for detailed discussion), verbal abuse and nonverbal rejection are reliable indicators of discriminatory effects, in that they disadvantage the targets of such behavior, creating a hostile environment. They also precede and vary with more overtly damaging forms of treatment, such as denial of employment (Dovidio et al., 2002; Fiske, 1998; Talaska et al., 2003). For example, an interviewer's initial bias on the basis of race will likely be communicated nonverbally to the interviewee by such behaviors as cutting the interview short or sitting so far away from the interviewee as to communicate immediate dislike (Darley and Fazio, 1980; Word et al., 1974). Such nonverbal hostility reliably undermines the performance of otherwise equivalent interviewees. In legal settings, verbal and nonverbal treatment are often presented as evidence of a discriminator's biased state of mind; they may also constitute unlawful discriminatory behavior when they rise to the level of creating a hostile work environment.

Avoidance entails choosing the comfort of one's own racial group (the "ingroup" in social psychological terms) over interaction with another racial group (the "outgroup"). In settings of discretionary contact—that is, in which people may choose to associate or not—members of disadvantaged racial groups may be isolated. In social situations, people may self-segregate along racial lines. In work settings, discretionary contact may force outgroup members into lower-status occupations (Johnson and Stafford, 1998) or undermine the careers of those excluded from informal networks.

Becker (1971) describes a classic theory about how aversion to interracial contact—referred to as a "taste for discrimination"—can affect wages and labor markets (more complex versions of this model are provided by Black, 1995; Borjas and Bronars, 1989; and Bowlus and Eckstein, 2002). Laboratory experiments have measured avoidance by assessing people's willingness to volunteer time together with an outgroup individual in a given setting (Talaska et al., 2003). Sociological studies have measured avoidance in discretionary social contact situations by report or observation (Pettigrew, 1998b; Pettigrew and Tropp, 2000). In legal settings, avoidance of casual contact can appear as evidence indicating hostile intent.

Avoidance may appear harmless in any given situation but, when cumulated across situations, can lead to long-term exclusion and segregation. It may be particularly problematic in situations in which social networking matters, such as employment hiring and promotion, educational opportunities, and access to health care. Avoiding another person because of race can be just as damaging as more active and direct abuse.

Segregation occurs when people actively exclude members of a disadvantaged racial group from the allocation of resources and from access to institutions. The most common examples include denial of equal education, housing, employment, and health care on the basis of race. The majority of Americans (about 90 percent in most current surveys; Bobo, 2001) support laws enforcing fair and equal opportunity in these areas. But the remaining

10 percent who do not support civil rights for all racial groups are likely to exhibit intentional, explicit discrimination by any measure. The data indicate that these hardcore discriminators view their own group as threatened by racial outgroups (Duckitt, 2001). They view that threat as both economic, in a zero-sum game, and as value based, in a contest of "traditional" values against nonconformist deviants. Moreover, even the 90 percent who report support for equal opportunity laws show less support when specific remedies are mentioned (see Chapter 8).

Physical attacks on racial outgroups have frequently been perpetrated by proponents of segregation (Green et al., 1999) and are correlated with other overt forms of discrimination (Schneider et al., 2000). Hate crimes are closely linked to the expression of explicit prejudice and result from perceived threats to the ingroup's economic standing and values (Glaser et al., 2002; Green et al., 1998; for a review of research on hate crimes, see Green et al., 2001).

Extermination or mass killings based on racial or ethnic animus do occur. These are complex phenomena; in addition to the sorts of individual hostility and prejudice described above, they typically encompass histories of institutionalized prejudice and discrimination, difficult life conditions, strong (and prejudiced) leadership, social support for hostile acts, and socialization that accepts explicit discrimination (Allport, 1954; Newman and Erber, 2002; Staub, 1989).

Our report focuses more on the levels of discrimination most often addressed by social scientists. In most cases involving complaints about racial discrimination in the United States, explicit discrimination is expressed through verbal and nonverbal antagonism and through racial avoidance and denial of certain opportunities because of race. Racial segregation is, of course, no longer legally sanctioned in the United States, although instances of de facto segregation continue to occur.

Subtle, Unconscious, Automatic Discrimination

Even as a national consensus has developed that explicit racial hostility is abhorrent, people may still hold prejudicial attitudes, stemming in part from past U.S. history of overt prejudice. Although prejudicial attitudes do not necessarily result in discriminatory behavior with adverse effects, the persistence of such attitudes can result in unconscious and subtle forms of racial discrimination in place of more explicit, direct hostility. Such *subtle prejudice* is often abetted by differential media portrayals of nonwhites versus whites, as well as de facto segregation in housing, education, and occupations.

The psychological literature on subtle prejudice describes this phenomenon as a set of often unconscious beliefs and associations that affect the attitudes and behaviors of members of the ingroup (e.g., non-Hispanic whites) toward members of the outgroup (e.g., blacks or other disadvantaged racial groups). Members of the ingroup face an internal conflict, resulting from the disconnect between the societal rejection of racist behaviors and the societal persistence of racist attitudes (Dovidio and Gaertner, 1986; Katz and Hass, 1988; McConahay, 1986). People's intentions may be good, but their racially biased cognitive categories and associations may persist. The result is a modern, subtle form of prejudice that goes underground so as not to conflict with antiracist norms while it continues to shape people's cognitive, affective, and behavioral responses. Subtle forms of racism are indirect, automatic, ambiguous, and ambivalent. We discuss each of these manifestations of subtle prejudice in turn (Fiske, 1998, 2002) and then examine their implications for discriminatory behavior.

Indirect prejudice leads ingroup members to blame the outgroup—the disadvantaged racial group—for their disadvantage (Hewstone et al., 2002; Pettigrew, 1998a). The blame takes a Catch-22 form: The outgroup members should try harder and not be lazy, but at the same time they should not impose themselves where they are not wanted. Such attitudes on the part of ingroup members are a manifestation of indirect prejudice. Differences between the ingroup and outgroup (linguistic, cultural, religious, sexual) are often exaggerated, so that outgroup members are portrayed as outsiders worthy of avoidance and exclusion. Indirect prejudice can also lead to support for policies that disadvantage nonwhites.

Subtle prejudice can also be unconscious and *automatic*, as ingroup members unconsciously categorize outgroup members on the basis of race, gender, and age (Fiske, 1998). People's millisecond reactions to outgroups can include primitive fear and anxiety responses in the brain (Hart et al., 2000; Phelps et al., 2000), negative stereotypic associations (Fazio and Olson, 2003), and discriminatory behavioral impulses (Bargh and Chartrand, 1999). People have been shown to respond to even subliminal exposure to outgroups in these automatic, uncontrollable ways (Dovidio et al., 1997; Greenwald and Banaji, 1995; Greenwald et al., 1998; Kawakami et al., 1998; for a review, see Fazio and Olson, 2003; for a demonstration of this effect, see https://implicit.harvard.edu/implicit/ [accessed December 5, 2003]). However, the social context in which people encounter an outgroup member can shape such instantaneous responses. Outgroup members who are familiar, subordinate, or unique do not elicit the same reactions as those who are unfamiliar, dominant, or undifferentiated (Devine, 2001; Fiske, 2002). Nevertheless, people's default automatic reactions to outgroup members represent unconscious prejudice that may be expressed nonverbally or lead to racial avoidance, which, in turn, may create a hostile, discriminatory environment. Such automatic reactions have also been shown to lead to automatic forms of stereotype-confirming behavior (Bargh et al., 1996; Chen and Bargh, 1997).

The main effect of subtle prejudice seems to be to favor the ingroup rather than to directly disadvantage the outgroup; in this sense, such prejudice is *ambiguous* rather than unambiguous. That is, the prejudice could indicate greater liking for the majority rather than greater disliking for the minority. As a practical matter, in a zero-sum setting, ingroup advantage often results in the same outcome as outgroup disadvantage but not always. Empirically, ingroup members spontaneously reward the ingroup, allocating discretionary resources to their own kind and thereby relatively disadvantaging the outgroup (Brewer and Brown, 1998). People spontaneously view their own ingroups (but not the outgroup) in a positive light, attributing its strengths to the essence of what makes a person part of the ingroup (genes being a major example). The outgroup's alleged defects are used to justify these behaviors. These ambiguous allocations and attributions constitute another subtle form of discrimination.

According to theories of ambivalent prejudice (e.g., for race, Katz and Hass, 1988; for gender, Glick and Fiske, 1996), the *ambivalence* of subtle prejudice means that outgroups are not necessarily subjected to uniform antipathy (Fiske et al., 2002). Outgroups may be disrespected but liked in a condescending manner. Versions of the "Uncle Tom" stereotype are a racial example. At other times, outgroups may be respected but disliked. White reactions to black professionals can exemplify this behavior. Some racial outgroups elicit both disrespect and dislike. Poor people, welfare recipients, and homeless people (all erroneously perceived to be black more often than white) frequently elicit an unambivalent and hostile response.

The important point is that reactions need not be entirely negative to foster discrimination. One might, for example, fail to promote someone on the basis of race, perceiving the person to be deferential, cooperative, and nice but essentially incompetent, whereas a comparable ingroup member might receive additional training or support to develop greater competence. Conversely, one might acknowledge an outgroup member's exceptional competence but fail to see the person as sociable and comfortable—therefore not fitting in, not "one of us"—and fail to promote the person as rapidly on that account.

All manifestations of subtle prejudice—indirect, automatic, ambiguous, and ambivalent—constitute barriers to full equality of treatment. Subtle prejudice is much more difficult to document than more overt forms, and its effects on discriminatory behavior are more difficult to capture. However, "subtle" does not mean trivial or inconsequential; subtle prejudice can result in major adverse effects.

For example, Bargh and colleagues (1996) demonstrated how categorization by race can activate stereotypes and lead to discriminatory behavior. In their study, the experimenter first showed white participants either black or white young male faces, presented at a subliminal level. The experimenter then either did or did not provoke the participant by requiring that the experiment be started over because of an apparent computer error. Compared with other participants, those who saw the black faces and were also provoked by the experimenter behaved with more hostility as revealed in a videotape of their immediate facial expressions and in their subsequent behavior, as rated by the experimenter.

Generally, an emerging pattern of results from laboratory research (see, e.g., Dovidio et al., 2002) suggests that explicit measures of prejudice (e.g., from responses to attitudinal questionnaires) predict explicit discrimination (verbal behavior), whereas implicit measures of prejudice (e.g., speed of stereotypic associations) predict subtle discrimination (such as nonverbal friendliness). In any event, the implicit measures have been shown to be statistically reliable (Cunningham et al., 2001; Kawakami and Dovidio, 2001).

Some of these laboratory findings have been generalized to the real world—for example, in contrasting subtle and explicit forms of prejudice (Pettigrew, 1998b) and in research on specific phenomena, such as ingroup favoritism (Brewer and Brown, 1998). The discussion of experimental methods in Chapter 6 elaborates on this point.

Statistical Discrimination and Profiling

Another process that may result in adverse discriminatory consequences for members of a disadvantaged racial group is known as *statistical discrimination* or *profiling*. In this situation, an individual or firm uses overall beliefs about a group to make decisions about an individual from that group (Arrow, 1973; Coate and Loury, 1993; Lundberg and Startz, 1983; Phelps, 1972). The perceived group characteristics are assumed to apply to the individual. Thus, if an employer believes people with criminal records will make unsatisfactory employees, believes that blacks, on average, are more likely to have criminal records compared with whites, and cannot directly verify an applicant's criminal history, the employer may judge a black job applicant on the basis of group averages rather than solely on the basis of his or her own qualifications.

When beliefs about a group are based on racial stereotypes resulting from explicit prejudice or on some of the more subtle forms of ingroup-versus-outgroup perceptual biases, then discrimination on the basis of such beliefs is indistinguishable from the explicit prejudice discussed above. Statistical discrimination or profiling, properly defined, refers to situations of discrimination on the basis of beliefs that reflect the actual distributions of characteristics of different groups. Even though such discrimination could be viewed as economically rational, it is illegal in such situations as hiring because it uses group characteristics to make decisions about individuals.

Why might employers or other decision makers employ statistical discrimination? There are incentives to statistically discriminate in situations in which information is limited, which is often the case. For example, graduate school applicants provide only a few pages of written information about themselves, job applicants are judged on the basis of a one-page resume or a brief interview, and airport security officers see only external appearance. In such situations, the decision maker must make assessments about a host of unknown factors, such as effort, intelligence, or intentions, based on highly limited observation.

Why is information limited in such cases? The decision maker typically views an individual's own statements about himself or herself as untrustworthy (e.g., "I will work hard on this job" or "I am not a terrorist") because they can be made as easily by those for whom they are not true as by those for whom they are true. Instead, decision makers look for signals that cannot easily be faked and are correlated with the attributes a decision maker is seeking. Education is a prime example. If an employer checks a job applicant's education credentials and finds that he or she has a degree from a top-rated college and a 4.0 grade point average, that individual likely has a proven track record of intellectual ability and effort. It is difficult to "fake" this information (short of outright lying about one's education credentials) because it really does take effort to accumulate such a record.

Only so much information can be transmitted, however, and many aspects of a person's record and qualifications are difficult to document even if the individual should be committed to doing so truthfully. Hence, decision makers must regularly make judgments about people based on the things they do know and decide whether to invest in acquiring further information (Lundberg, 1991). In the face of incomplete information, they may factor in knowledge about differences in average group characteristics that relate to the individual characteristics being sought. The result is statistical discrimination: An individual is treated differently because of information associated with his or her racial group membership.

Faced with the possibility of statistical discrimination, members of disadvantaged racial groups may adopt behaviors to signal their differences from group averages. For example, nonwhite business people who want to signal their trustworthiness and belonging to the world of business may dress impeccably in expensive business suits. Nonwhite parents who want their children to get into a first-rate college may signal their middle-class background by sending their children to an expensive private school. An implication of statistical discrimination is that members of a disadvantaged racial group for whom group averages regarding qualifications are lower than white averages may need to become better qualified than non-Hispanic whites in order to succeed (Biernat and Kobrynowicz, 1997). Thus, the practice of statistical discrimination can impose costs on members of the targeted group even when those individuals are not themselves the victims of explicitly discriminatory treatment.

Moreover, statistical discrimination may be self-perpetuating, since today's outcomes may affect the incentives for tomorrow's behavior (Coate and Loury, 1993; Loury, 1977; Lundberg and Startz, 1998). If admissions officers at top-ranked colleges believe, on the basis of group averages to date, that certain groups are less likely to succeed and admit few members of those groups as a result, incentives for the next generation to work hard and acquire the skills necessary to gain admittance may be lessened (see Loury, 2002:32–33, for a more extensive discussion of this example). Similarly, if black Americans are barred from top corporate jobs, the incentives for younger black men and women to pursue the educational credentials and career experience that lead to top corporate jobs may be reduced. Thus, statistical discrimination may result in an individual member of the disadvantaged group being treated in a way that does not focus on his or her own capabilities. It can affect both short-term outcomes and long-term behavior if individuals in the disadvantaged group expect such discrimination will occur.

Organizational Processes

The above three types of racial discrimination focus on individual behaviors that lead to adverse outcomes and perpetuate differences in outcomes for members of disadvantaged racial groups. These behaviors are also the focus of much of the current discrimination law. However, they do not constitute a fully adequate description of all forms of racial discrimination. As discussed in Chapter 2, the United States has a long history as a racially biased society. This history has done more than change individual cognitive responses; it has also deeply affected institutional processes. Organizations tend to reflect many of the same biases as the people who operate within them. Organizational rules sometime evolve out of past histories (including past histories of racism) that are not easily reconstructed, and such rules may appear quite neutral on the surface. But if these processes function in a way that leads to differential racial treatment or produces differential racial outcomes, the results can be discriminatory. Such an embedded institutional process—which can occur formally and informally within society—is sometimes referred to as *structural discrimination* (e.g., Lieberman, 1998; Sidanius and Pratto, 1999). In Chapter 11, we discuss the interactions among these processes that occur within and across domains.

One clear example of this phenomenon occurs in the arena of housing.

In the past, overt racism and explicit exclusionary laws promoted residential segregation. Even though these laws have been struck down, the process by which housing is advertised and housing choices are made may continue to perpetuate racial segregation in some instances. Thus, real estate agents may engage in subtle forms of racial steering (i.e., housing seekers being shown units in certain neighborhoods and not in others), believing that they are best serving the interests of both their white and their nonwhite clients and not intending to do racial harm. Likewise, banks and other lending institutions have a variety of apparently neutral rules regarding mortgage approvals that too often result in a higher level of loan refusals for persons in lower-income black neighborhoods than for equivalent white applicants. Research also suggests that ostensibly neutral criteria are often applied selectively. Credit history irregularities that are overlooked as atypical in the case of white mortgage applicants, for example, are often used to disqualify blacks and Latinos (Squires, 1994; Squires and O'Connor, 2001).

Another example of this sort of biased institutional process that has been debated in the courts is the operation of hiring and promotion networks within firms. Many firms hire more through word-of-mouth recommendations from their existing employees than through external advertising (Waldinger and Lichter, 2003). By itself such a practice is racially neutral, but if existing (white) employees recommend their friends and neighbors, new hires will replicate the racial patterns in the firm, systematically excluding nonwhites. Such practices do not necessarily entail intentional discrimination, but they provide a basis for legal action when the outcome is the exclusion of certain groups. Seniority systems that give preference to a long-established group of employees can produce similar racially biased effects through promotion or layoff decisions, even though the Supreme Court has ruled that seniority systems are generally not subject to challenge under Title VII on this basis.[1]

Institutional processes that result in consistent racial biases in terms of who is included or excluded can be difficult to disentangle. In many cases, the individuals involved in making decisions within these institutions will honestly deny any intent to discriminate. In dealing with such cases in the courts (disparate impact cases; see Chapter 3), weighing the benefits to an organization of a long-established set of procedures against the harm such procedures might induce through their differential racial outcomes is a complex and difficult process. Thus the panel does not wish to condemn any specific organizational process. In most cases, each situation needs to be

[1]International Brotherhood of Teamsters v. United States, 431 U.S. 324 (1977) (the "routine application of a bona fide seniority system" is not unlawful under Title VII).

analyzed with regard to the particular history and reasonable organizational needs of a specific institution. But we do want to emphasize that facially neutral organizational processes may function in ways that can be viewed as discriminatory, particularly if differential racial outcomes are insufficiently justified by the benefits to the organization. We noted above that large and persistent racial differentials, although not direct evidence of discrimination, may provide insight on where problems are likely to exist. In this way, persistent racial differences in access to or outcomes within institutions (e.g., hiring or promotions) can be used to provide information on which processes and which institutions may deserve greater scrutiny.

COMPARISON OF LEGAL STANDARDS WITH THE FOUR TYPES OF DISCRIMINATION

As discussed in Chapter 3, the legal definition of discrimination includes two standards: disparate treatment discrimination, whereby an individual is treated less favorably because of race, and disparate impact discrimination, whereby treatment on the basis of nonracial factors that lack sufficiently compelling justification has an adverse impact on members of a disadvantaged racial group. The quintessential case of disparate treatment discrimination involves intentional behavior motivated by explicit racial animus. However, disparate treatment applies in other types of discrimination as well. For instance, a black cab driver who refuses to pick up blacks may be acting without racial animus but may be engaging in statistical discrimination by making probabilistic predictions about the risk of being victimized by crime, of receiving a lower tip, or of ending up in a distant neighborhood from which the prospect of receiving a return fare is small. Employers and police officers who profile job candidates or security risks can be motivated by similar beliefs or concerns, and their probabilistic assessments may be correct or completely inaccurate. In any event, as noted above, this type of statistical discrimination is considered intentional differentiation on the basis of race and falls squarely in the category of unlawful disparate treatment discrimination. In evaluating a job applicant, for example, it is unlawful to consider what the "average" black worker would be like and then to treat individual blacks in conformity with this stereotypical prediction.

In short, although vexing issues of proof complicate real-world cases, the law has clearly identified the theoretically prohibited discriminatory actions that emanate from either racial animus or the rational calculation of risk using race as a proxy. More subtle types of discrimination, however, are more difficult to deal with legally. As discussed above, there may be no conscious bias or rational calculation that prompts someone to treat whites differently from nonwhites. Such precognitive patterns of conduct have been well documented and are in practice treated as cases of unlawful disparate treatment discrimination if they are found to generate differential treatment of blacks. Note, however, that issues of proof make it more difficult to establish these unconscious forms of discriminatory behavior, although statistical approaches are commonly used to ferret out just such unconscious bias. Indeed, the legal requirement that unlawful disparate treatment discrimination must involve intentional discrimination may result in many indirect, subtle, and ambiguous types of discrimination being overlooked. In some cases, nonetheless, an organization has been found guilty of intentional discrimination for failing to compensate for the unconscious, automatic discrimination of its employees.

DOMAINS IN WHICH DISCRIMINATION OPERATES

As discussed in Chapter 1, this report focuses on the measurement of discrimination in specific domains: labor markets and employment, education, housing and mortgage lending, criminal justice, and health care. The focus on these areas reflects the expertise of the members of this panel. There are a variety of other domains, such as civic participation, in which racial differences in outcomes are large, and discrimination is a valid social concern. We believe that our comments about assessing discrimination, although directed at the domains and examples with which we are most familiar, may be useful and applicable in other arenas as well. In this section, we briefly review some of the key points at which the forms of discrimination delineated above may operate within the domains on which we focus.

Table 4-1 shows how discrimination might operate across the five domains of labor markets, education, housing, criminal justice, and health care at three broadly defined points. The first point is discrimination in access to the institutions within a domain; examples are racial differentials in hiring in the labor market, racial steering in housing, financial aid for schooling, arrest rates or policing activity within communities, and access to certain medical institutions or procedures. The second point is discrimination while functioning within a domain; examples are racial differentials in wages, mortgage loan pricing, placement into special education programs, assignment of pro bono legal counsel, and quality of health care. Closely related is discrimination in movement or while progressing within a domain from one activity to another; examples are racial differentials in job promotions, home resale value, grade promotion in schools, sentencing or parole rates, and medical referrals or follow-up health care. Of course, such discrimination often follows discriminatory behavior at an earlier point in time. Finally, the table lists possible actors within each domain who may discriminate on the basis of race. These actors include employers, customers, and coworkers in the labor market; teachers, administrators, and students

TABLE 4-1 A Map of the Potential Points of Discrimination Within Five Domains

Source Points for Discrimination	Labor Markets	Education	Housing/ Mortgage Lending	Criminal Justice	Health Care
Access to institutions or procedures	• Hiring • Interviewing • Unemployment	• Acceptance —Into college —Into special education programs • Financial aid	• Steering • Mortgage redlining	• Policing behaviors • Arrests	• Access to care • Insurance
While functioning within a domain	• Wages • Evaluation • Work environment	• Track placement • Ability grouping • Grades and evaluations • Learning environment • Per-pupil expenditure • Special education placement	• Loan pricing	• Police treatment • Quality of legal representation	• Quality of care • Price
Movement through a domain	• Promotion • Layoffs • Rehiring	• Promotion and graduation • Retention	• Resale value • Wealth accumulation	• Parole • Sentencing	• Referrals
Key actors	• Employers • Customers • Coworkers	• Teachers • Administrators • Fellow students	• Landlords • Sellers • Lenders • Neighbors	• Police • Prosecutors • Judges • Juries • Parole boards	• Health care workers • Administrators • Insurance companies

NOTE: We provide a selected bibliography of research on discrimination within the domains listed above at the end of this report.

in schools; landlords, sellers, lenders, and neighbors in housing; police officers, judges, and juries in criminal justice; and health care professionals, insurance companies, and administrators in the health care system.

At any of the points shown in the table, one might observe direct adverse behavior or aversion to contact with racial minorities, unconscious or subtle biases, statistical discrimination, or institutional processes that result in adverse outcomes. The remainder of this report addresses the methods that are used to investigate possibly discriminatory behavior within the various cells of this matrix.

We do not attempt to provide a comprehensive review of the literature on racial discrimination within each of the categories and domains listed in Table 4-1. Several extensive articles and reports review the literature within specific domains. We provide a selected bibliography of major papers from the theoretical and empirical literature at the end of this report. This bibliography includes research that demonstrates the methods used to assess discrimination within particular domains. Although in Part II of our report we do not discuss specific methods applied in each domain in turn, we do examine the broad approaches used to measure the types of discrimination outlined above. We also discuss where alternative approaches may be implemented more easily within one domain than another. In some cases, we suggest that specific methods should be applied in domains where they have not yet been used.

MOVING FROM EPISODIC TO DYNAMIC DEFINITIONS OF DISCRIMINATION: THE ROLE OF CUMULATIVE DISADVANTAGE

Much of the discussion of the presence of discrimination and the effects of antidiscrimination policies assumes discrimination is a phenomenon that occurs at a specific point in time within a particular domain. For instance, discrimination can occur in entry-level hiring in the labor market or in loan applications in mortgage lending. But this episodic view of discrimination occurring may be inadequate. Here we explore the idea, noted in Chapter 3, that discrimination should be seen as a dynamic process that functions over time in several different ways.

First, the effects of discrimination may cumulate across generations and through history. For instance, impoverishment in previous generations can prevent the accumulation of wealth in future generations. Similarly, learned behavior and expectations about opportunities and life possibilities can shape the behaviors and preferences of future generations for members of different racial groups.

Second, effects of discrimination may cumulate over time through the course of an individual's life across different domains. Outcomes in labor markets, education, housing, criminal justice, and health care all interact with each other; discrimination in any one domain can limit opportunities and cumulatively worsen life chances in another. For instance, children who are less healthy and more impoverished may do worse in school, and in turn, poor education may affect labor market opportunities. The possibility that the effects of discrimination cumulate over an individual's lifetime is rarely discussed in the literature on the measurement of discrimination. Yet even small initial disadvantages, experienced at key points in an individual's life, could well have long-term cumulative effects.

Third, effects of discrimination may cumulate over time through the course of an individual's life sequentially within any one domain. Again, small levels of discrimination at multiple points in a process may result in large cumulative disadvantage. For instance, children who do not learn basic educational skills in elementary school because of discrimination may face future discrimination in the way they are tracked or the way their test scores are interpreted in secondary school. Small effects of discrimination in job search (e.g., application or interviewing stages), job retention, job promotion, and wage setting may result in large differences in labor market outcomes when these effects cumulate over time, even if no further discrimination occurs.

There are many instances in which the application of neutral rules harms a member of a disadvantaged racial group because of discrimination at some other time or place in the social system. However, there is presently no case law that addresses these broad social effects; the law frequently will not deem the challenged conduct to be unlawful if it merely transmits, rather than expands, the extent of racial discrimination. Similarly, the law does not hold any agents or institutions responsible for problems outside their legitimate purview. Discrimination occurring in other domains or in society generally need not be remedied; hence, cumulative discrimination is not a legal issue. An employer who needs highly educated workers can hire them as he or she finds them, even if doing so means that only a small percentage of black or Hispanic workers will be hired because prior discrimination in educational opportunities limited the number of members of these groups with the requisite skills.

Whether cumulative discrimination is important across generations, across a lifetime in different domains, and over time within a specific domain are empirical questions. However, these questions have not been addressed to any great extent by empirical social scientists. In Chapter 11, we return to the issue of the importance of developing methods focused not just on measuring discriminatory behavior at a particular point in time in a specific process but also on understanding the cumulative and dynamic effects of discrimination over time and across processes.

SUMMARY

Discrimination manifests itself in multiple ways that range in form from overt and intentional to subtle and ambiguous, as well as from personal to institutional, whether through statistical discrimination and profiling or organizational processes. Discrimination also operates differently in different domains and may cumulate over time within and across domains. Regardless of which form it takes, discrimination can create barriers to equal treatment and opportunity and can have adverse effects on various outcomes. Clear theories about how discriminatory behavior may occur are important in order to develop models that help identify and measure discrimination's effects.

Although discrimination is sometimes still practiced openly, it has become increasingly socially undesirable to do so. Consequently, such discrimination as exists today is more likely to take more subtle and complex forms. Subtler forms of discrimination can occur spontaneously and ambiguously and go undetected, particularly at the institutional level. Although legal standards address specific forms of unlawful intentional or statistical discrimination, subtler forms are more difficult to address within the law. Thus, shifts in kinds of discriminatory behavior have implications for the measurement of discrimination. As we discuss in the next chapter, some types of discrimination may be more difficult to identify and may require collecting new and different data and the further development of new methods of analysis.

Part II

Methods

In Part I, we defined the concepts of race and racial discrimination from a social science research perspective. The history of legal and institutionalized racial discrimination in the United States and the existence of widespread racial disparities in outcomes across domains prompt our review of methods for assessing the extent to which discrimination continues to affect historically disadvantaged racial and ethnic groups. Our definition of racial discrimination includes overt and subtle discriminatory behaviors and processes. If, as some have suggested, modern forms of discrimination are less likely to be direct and explicit and more likely to be indirect and ambiguous than in the past, it will be increasingly difficult to measure the effects of discrimination on various outcomes.

Our goal in Part II is to move from a descriptive analysis of existing disparities (association) to consider methods of inferential analysis (causation), with a focus on determining the circumstances in which a racial disparity may be attributed, in whole or in part, to racial discrimination. The core measurement issues in which we are interested include the following:

- measuring the incidence, causes, and effects of racial discrimination;
- identifying appropriate units of analysis (individual or aggregate level);
- identifying explanatory mechanisms that lead to discriminatory behaviors and institutional processes;
- identifying mediating factors and processes that affect observed disparities;
- measuring the extent or magnitude of discrimination within a do-

main, across domains, and over time; and

- determining how much of an observed disparity is an effect of discrimination.

Our discussion of these issues is limited by our charge to focus on the measurement of racial discrimination. However, much of the discussion can be readily applied to measurement of closely related topics, such as gender or age discrimination.

There are many different methods for measuring racial discrimination. We review three types of methods that are widely used in various literatures: controlled laboratory experiments and field experiments; analysis of observational data and natural experiments; and measures of reported perceptions and experiences of discrimination from surveys and administrative records. It is important to note that no one method allows researchers to address all of the measurement issues listed above.

For example, laboratory experiments help researchers to identify the mechanisms that may lead to different forms of racial discrimination and the factors that mediate the expression of discriminatory attitudes and behaviors. Because of experimental control over relevant variables, researchers are able to identify whether race or the interaction of race and other factors triggers an expression of racial discrimination. Laboratory experiments are useful for drawing causal inferences at the individual level and important for identifying subtle mechanisms of discrimination; however, they do not directly address disparities in the aggregate. That is, laboratory effects do not often generalize to the broader population and can rarely tell us the extent to which naturally observed disparities are the result of discrimination.

The results of field experiments, on the other hand, are often more generalizable than the results of laboratory experiments. Although field experiments may involve less experimental control, researchers can use them to measure the extent of discrimination in a particular domain, such as the housing or labor market. For instance, audit studies in the housing or employment arena can provide useful information about the possible occurrence of discrimination by real estate agents against homebuyers or by employers against job applicants from disadvantaged racial groups.

Some ability to generalize may also be gained by using nonexperimental approaches. Researchers can use statistical modeling and estimation to analyze observational data and draw causal inferences. Statistical models are useful for identifying associations between race and different outcomes while controlling for other factors that may explain the observed outcomes. Simply identifying an association with race, however, is not equivalent to measuring the magnitude of racial discrimination or its contribution to differential outcomes by race. In most observational settings, the lack of experimental control and the inability to manipulate "treatment" variables make it difficult to dismiss alternative explanations of causation without relying on strong and often untestable assumptions.

One problem is that observational data often contain only a small set of characteristics and may not include variables that are important for explaining an observed effect or for modeling the process by which discrimination could occur. For instance, a finding of a large discriminatory effect within a domain (e.g., differential treatment in hiring leading to wage disparities) may be erroneous if it is not possible to control for other explanatory variables, such as motivation or skill level,[1] or to develop an accurate statistical model of the decision process. Alternatively, a finding that discrimination at a certain point within a domain contributes little to an observed disparity may ignore the possible effects of how earlier discrimination may have accumulated over time; for example, discrimination in secondary education can affect skill levels and thereby affect subsequent wages (see Chapters 4 and 11).

Surveys also provide observational data to measure racial attitudes and reported experiences and perceptions of racial discrimination. But again, these data are rarely sufficient to establish causality, statistically or substantively. The most detailed observational studies are collections of case studies, which contain large amounts of information on small numbers of individuals or organizations. Such collections of case studies can produce the kinds of information on underlying behavioral processes needed to draw valid causal inferences, although their results may be limited in generalizability. Longitudinal survey data can be particularly helpful for understanding trends in racial attitudes and reported experiences and perceptions of discrimination and the extent to which racial disparities are a function of discrimination that occurs over time and across domains.

Determining how much of an observed outcome is an effect of racial discrimination is difficult. Translating effects from experimental data to what is observed in real situations is not easy. Moreover, it is much easier to assess the occurrence of discrimination at one point in a process than to identify effects of discrimination that occur earlier in a process or across relevant domains. A feasible solution to these difficulties may be to combine methods, using data and results from multiple sources. In the following five chapters, we describe issues and methods for research design, measurement, and analysis that together may allow researchers to identify and assess racial discrimination. When appropriate, each chapter contains conclusions and recommendations.

---

[1]Goldsmith et al. (2000) provide a counterexample of research that explicitly includes measures of motivation.

We begin in Chapter 5 by introducing a general framework for inferring causation between race-based discrimination and outcomes of interest. Racial disparities are often substantial and widely observed, but only rarely do researchers directly observe discriminatory behavior. To establish a causal relationship between race and discrimination, one would ideally vary the race of a single person and measure any differences in outcomes. Because doing so is impossible, researchers typically observe a disparate outcome and trace back through the process that generated it to determine whether racial discrimination had a causal effect. In other words, they attempt to answer retrospectively the counterfactual question of whether the outcome for a nonwhite individual would have been different if he or she had been white. In Chapter 5, we discuss how accumulated scientific evidence from both experimental and observational research may support causal conclusions and allow researchers to determine whether racial discrimination contributes causally to an observed racial disparity.

Controlled experiments, whose strengths are direct manipulation of experimental conditions and randomization, are ideal for drawing causal inferences and come closest to addressing the above counterfactual question. Chapter 6 describes two types of experimental methods—laboratory and field experiments—and their application to the measurement of racial discrimination. We describe the design, use, strengths, and limitations of experimental methods and provide key examples of laboratory and field studies used to measure racial discrimination.

In Chapter 7, we critically review the issues that must be addressed to draw valid causal inferences about racial discrimination from analyses of observational data. We first review the primary descriptive approach used in the literature on racial discrimination—decomposition of racial disparities. We then discuss the limitations of such descriptive analysis and the challenges of moving from description to inference by using statistical models (particularly regression models). We focus on the assumptions underlying statistical models and possible approaches to the problems of using such models to infer discrimination, one of which is to take advantage of natural experiments, which occur when a policy change targets discrimination in a particular domain or set of domains.

Our primary intent in Chapter 8 is to consider the use of observational data from surveys, in-depth interviews, and administrative records to measure the occurrence of racially discriminatory attitudes and behaviors and people's reported perceptions of and experiences with discrimination. We provide extended reviews of major large-scale surveys of racial attitudes and reported perceptions and experiences of racial discrimination by white and black Americans.

Finally, in Chapter 9 we provide an example of the challenges of measuring racial discrimination in an important area of current concern, which is racial profiling by law enforcement officials in which race alone or in combination with other variables is used to select individuals for further investigation (e.g., to stop motorists to search for illegal drugs). Profiling is a form of statistical discrimination. We discuss some of the methodological challenges of determining when racial profiling may be occurring, although these challenges are such that we are not able to identify the best measurement approaches.

5

Causal Inference and the Assessment of Racial Discrimination

Because discriminatory behavior can rarely be directly observed, researchers face the challenge of determining when racial discrimination has actually occurred and whether it explains some portion of a racially disparate outcome. Those who attempt to identify the presence or absence of discrimination typically observe an individual's race (e.g., black) and a particular outcome (e.g., earnings) and try to determine whether that outcome would have been different had the individual been of a different race (e.g., white). In other words, to measure discrimination researchers must answer the counterfactual question: What would have happened to a nonwhite individual if he or she had been white? Answering this question is fundamental to being able to conclude that there is a causal relationship between race and discrimination, which, in turn, is necessary to conclude that race-based discriminatory behaviors or processes contributed to an observed differential outcome.

To illustrate the problem, we turn to a classic Dr. Seuss book, *The Sneetches* (published in 1961), which describes a society of two races distinguished by markings on their bellies. In the story, one race of Sneetches is afforded certain privileges for having stars on their bellies, and the other race, lacking these markings, is denied those same privileges. There are, however, Star-On and Star-Off machines that can alter the belly and therefore the race of both Plain-Belly and Star-Belly Sneetches. Thanks to these machines, an individual Sneetch's racial status and various outcomes could be observed more than once, both as a Plain-Belly and a Star-Belly Sneetch.

In *The Sneetches*, belly-based discrimination is evident in the society; the causal relationship between race and discrimination can be ascertained because stars can be placed on or removed from any belly by a machine, and multiple outcomes can be observed for a single Sneetch. Therefore, one could readily answer the counterfactual question, saying with certainty what would have happened to a Plain-Belly Sneetch had he or she been a Star-Belly Sneetch (or vice versa).[1] The phenomenon of a black individual passing as white (or vice versa) is an example of how race can be manipulated in this way in our society; thus, it is potentially interpretable causally. However, almost all the information on passing is anecdotal and there are few attempts to measure it systematically. Except under these circumstances, it is nearly impossible in the real world to observe the difference in outcomes across race for a single person; one must instead draw causal inferences.

DRAWING CAUSAL INFERENCES

In the context of measuring racial discrimination, researchers have developed alternative methods to answer the above counterfactual question and assess the incidence and effects of racial discrimination. A formal account of the counterfactual approach to causal inference provides a foundation for evaluating alternative solutions.

Counterfactuals and Potential Outcomes

Counterfactual analysis, combining elements of counterfactual and manipulability theories, is the dominant causal paradigm in recent literature in statistics. The past two decades have witnessed a growing literature formalizing the assumptions and the deductive process needed to draw cause-and-effect inferences from statistical data (Freedman, 2003; Holland, 1986, 2003; Pearl, 2000; Pratt and Schlaifer, 1984, 1988; Rubin, 1974, 1977, 1978; Spirtes et al., 1993; see Box 5-1 for a discussion of graphical approaches). The counterfactual approach to causal inference underlies work in sociology, appearing in both methodological discourse and substantive applications (see Gamoran and Mare, 1989; Lucas and Gamoran, 1991, 2002; Morgan, 2001; Sobel, 1995, 1996; Winship and Morgan,

[1]This example, while nicely illustrating our methodological point about causality, overlooks a key point. A world in which individuals could change their race as readily as Dr. Seuss's Sneetches can add or remove the stars on their bellies would be a world in which deeply ingrained racial inequalities could not exist. Members of a disadvantaged group would merely exercise the option to join the privileged group. Although there are accounts of individuals "passing" as a different race (such as depicted in John Howard Griffin's 1996 book *Black Like Me*), we generally do not live in such a world. Hence, the virtual immutability of race at the individual level is not only a barrier to drawing causal inferences about discrimination but also a necessary condition for the existence of a racial hierarchy in the first place.

1999; Winship and Sobel, 2004). Central to such cause-and-effect inferences is the notion of the manipulability of the potential causal variable, such as race. Because race cannot be directly manipulated or randomly allocated to study participants, researchers must be able to translate experimental results into a framework that allows them to address, in some form, causal statements regarding evidence of discrimination. Holland (2003) makes this point in detail and reiterates his earlier (1986) argument that one cannot have causation without manipulation. However, Holland also argues for the careful study of the interactions of race with manipulable variables. See Marini and Singer (1988) for a different perspective on causation.

As suggested above, causal questions are counterfactual questions. The causal effect of racial discrimination is the difference between two outcomes: the outcome if the individual were black and the outcome if the individual were white.[2] Rubin (1974) describes the fundamental problem—the inability to simultaneously observe different outcomes for the same person—as a missing data problem: Each individual has potential outcomes under each set of circumstances, but only one of these outcomes is observed (or realized). The causal effect of interest is the difference between these potential outcomes—that is, the effect of racial discrimination. The entire enterprise of causal inference is centered on alternative approaches for overcoming our inability to observe both of these outcomes for a single individual.[3]

Imagine we want to estimate the effect of discrimination on earnings as experienced by a black person. At the individual level, the unit causal effect of racial discrimination (here, discrimination against a black individual) is

$$Y_b - Y_w,$$

where Y_b represents the black individual's potential earnings, and Y_w repre-

[2]To be strictly consistent with the traditional literature on causal inference, we would call this difference the causal effect of race. However, the effect in which we are interested is the effect of race-based discriminatory behavior. We interpret a nonnegligible "effect of race," in this context, as indicative of racial discrimination.

[3]According to a conservative statistical position articulated by Freedman (2003) and others, we cannot draw *any* causal inferences in the absence of manipulability. Thus, viewed as a nonmanipulable attribute, race cannot be said to have a causal effect. Others have suggested that by considering the manipulation of all possible confounders, we can at least create a framework in which causal statements about nonmanipulable variables such as race are possible. This latter position is worth serious explication in the context of the measurement of discrimination and is related to ideas set forth in the economics literature going back to Havelmo in the 1940s. A full exploration of this position is beyond the scope of this study, however.

BOX 5-1
Illustrating Causality

Over the past 15 years, directed acyclic graphs have been introduced into the statistical and philosophical literature to describe statistical models and the causal relationships they capture. In this type of graph, each node represents a separate variable, and it is important for the graph to include unobserved variables that influence observable variables. Directed edges between nodes represent causal relationships between variables. By definition, paths following the directed edges in an acyclic graph cannot lead from a node back to itself. In the formal statistical theory of directed acyclic graphs (Pearl, 2000), the absence of an edge in the graph corresponds to conditional independence of the variables corresponding to the nodes, given all of the other variables represented in the graph. This use of conditional independence allows the tie-in to the formal structure for causal inference we have just described.

These diagrams offer a way to visualize causal relationships and the role of counterfactuals. In such graphs, manipulation in the sense we have described changes the graph by severing the links to other variables in the graph and, when done using formal randomization, adds a new random variable to the graph that breaks the link between the possible cause X and all of the other variables except for the outcome variable Y. Thus, we can conclude from the altered graph that X indeed causes Y, and we have a justification for the use of the experimental data to estimate the quantity in the text: $\alpha = E(Y \mid X = x_1) - E(Y \mid X = x_2)$. Figure 5-1 depicts two such graphs in accordance with the general framework presented here. Graph A depicts the causal relationships in the observational setting; graph B depicts the causal relationships under randomized assignment. This is merely a graphical representation of the process described above, where X is the treatment or cause of interest, Z is a vector

sents the potential earnings if the same individual were white.[4] However, we observe only Y, one of the two potential outcomes (Y_b or Y_w), depending on whether the individual is, in fact, black or white. To draw a causal inference about the incidence and effect of racial discrimination, the researcher reframes the question at the population level and then exploits

[4]Note that a causal effect must always be couched in terms of an alternative treatment or control—white or black, blonde or brunette, and so on. A causal effect is defined by both the "cause" of interest and its alternative. In the statistical literature on discrimination, the two alternatives are typically expressed in terms of expectations of the random variables Y_b and Y_w for relevant populations or distributions of individuals.

of observed covariates, U is a vector of unobserved covariates, and Y is the outcome. The introduction of random assignment, R, eliminates the directed edges from Z and U to X.

The counterfactual representation of causal probability statements has been an integral part of this new literature on directed acyclic graphs, although there have been recent attempts to use the graphical framework that do not explicitly incorporate counterfactuals (e.g., see Dawid, 2000). However, these efforts retain manipulation as a key element.

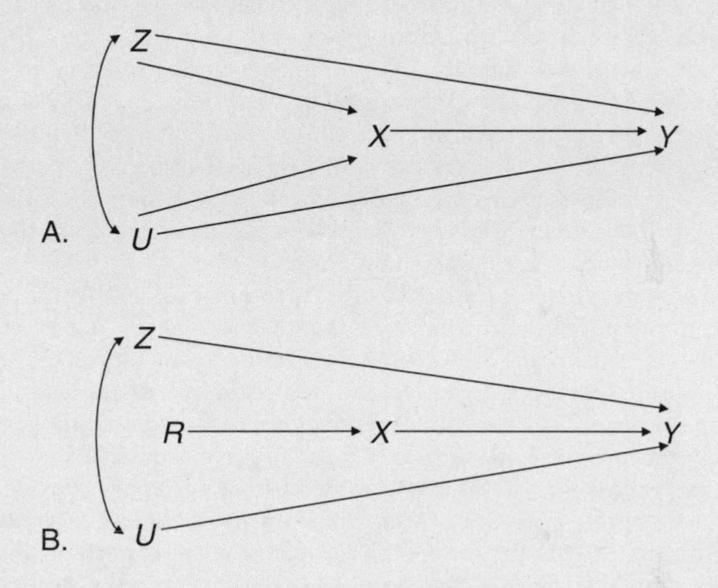

FIGURE 5-1 Directed acyclic graphs to depict causal relationships.

knowledge of population averages of outcomes among aggregates of members of a racial group to estimate the average effect of racial discrimination.

Study Design and Statistical Methods

Research design is critical to the ability to draw causal inferences from data analysis. For purposes of causal inference, there is a hierarchy of approaches to data collection. As one moves from meticulously designed and executed laboratory experiments through the variety of studies based on observational data, increasingly strong assumptions are needed to support the claim that X "causes" Y. The more careful and rigorous the design and control, the stronger are the inferences that can be drawn, provided that the design and control are used to address the causal question of interest.

Experimental Designs

The randomized controlled experiment is typically found at the top of the hierarchy of methodological approaches in terms of rigor and control. Such experiments involve direct manipulation of experimental treatments and random assignment of participants to treatments, which is believed to result in a balancing over unmeasured (and sometimes measured) variables whose effects must be controlled for if one is to infer causation. (We provide a more extensive discussion of experiments related to racial discrimination, as well as a variety of examples of such research, in Chapter 6.) Controlled experiments have *internal validity* associated with inferences about causation for the units of study in the experiment (i.e., the participant sample). But additional or extra-experimental information is required to achieve *external validity*, whereby researchers can generalize from the units in the experiment to some larger population.

One way to achieve external validity is to draw experimental participants from the population of interest. Another is to carry out a series of replications designed to allow for generalization from the set of experiments. Many researchers argue for the "universality" of the causal phenomenon measured in their experiment (i.e., the effect holds more generally, independently of both the other variables and context), but there is serious uncertainty about this claim in the absence of replications or other extra-experimental information. Many experimenters also argue for the role of experiments as demonstrations of the plausibility of particular causal processes, that is, as an existence proof that a particular phenomenon can, under at least some circumstances, occur in a particular manner. But this demonstration of plausibility does not address the issue of external validity. In any well-designed and well-executed experiment, randomization allows researchers to dismiss competing explanations as highly unlikely, but they are not entirely eliminated. For this reason, independent replication is important.

Holland (2003), in addressing related issues of causation and race, attempts to distinguish among three types of causal questions: (1) identifying causes, (2) assessing effects, and (3) describing mechanisms. Identifying causes is often a form of speculative postmortem. Randomized experiments are used to assess effects, but Holland argues that they can rarely be used to measure the effects of discrimination. On the other hand, Holland's notion of describing mechanisms relates to what this report refers to as understanding the process whereby discrimination may be occurring. He argues that to conduct an experiment one does not need to fully describe mechanisms. This report views the effort to measure the unobserved counterfactual usually associated with experiments as necessarily being linked to a detailed understanding of the process.

Observational Studies

Moving down the hierarchy with regard to rigor, especially for causal inferences, there are several intermediate steps between conducting controlled experiments and simply observing an event in an unstructured way. Nonexperimental methods differ from those used in experiments in that the analyst cannot assign particular (racial) attributes to particular subjects when in a nonexperimental setting. In observational studies, researchers have data on units at a point in time in a one-time sample survey, or longitudinally in a multiwave sample survey, or from another source (e.g., detailed case studies).

The available data may provide reports of perceived experiences of discrimination and discriminatory attitudes. Such data, obtained using random sampling and exhibiting low nonresponse and error rates, may allow for the external validity or generalizability that many experiments lack. However, because such survey-based studies measure subjective reports, they can be used to investigate only a limited number of phenomena related to but not measuring discrimination, such as trends in overtly discriminatory attitudes or in perceived discriminatory events (see discussion in Chapter 8).

Alternatively, the available data may provide information on differential outcomes (e.g., wage rates) for racial groups together with other variables that the researcher may use to infer the possible role of race-based discrimination. In such passive observation, the researcher lacks control over the assignment of treatments to subjects and attempts to compensate for this lack by "statistically controlling" for possible confounding variables (we elaborate on this issue in Chapter 7). In such circumstances, causal inferences can be controversial.

Statistical methods developed for drawing causal inferences are organized largely around trying to re-create, from observational data, the circumstances that would have occurred had controlled experimental data been collected. These statistical methods are discussed in some detail in Chapter 7, where we critically review the use of statistical models, particularly regression models, to draw valid causal inferences from observational data.

The Roles of Randomization and Manipulation

Researchers justify the substitution of population-level expectations for individual-level outcomes by designing experiments that incorporate elements of randomization and manipulability. Experiments allow us to distinguish more readily between prediction (or association) and causation (or intervention or manipulation). To predict a random variable Y from the variable X, we attempt to measure or estimate expected values or probabilities like

$$\Pr(Y = y \mid \text{Observe } X = x),$$

which is the probability that an outcome Y is equal to y, given that we have observed that some characteristic X is equal to x (e.g., lung cancer is more frequent among cigarette smokers). In contrast, to infer that smoking cigarettes *causes* an increase in the risk of lung cancer, we attempt to measure or estimate

$$\Pr(Y = y \mid \text{Set } X = x),$$

which is the probability that an outcome Y is equal to y, given that we have an assigned value of x for the random variable X. That is, in experiments designed to demonstrate causation, researchers manipulate X, setting its value for each experimental unit. By using different values of the quantity X for different units in the study, researchers are able to compare outcomes conditional on each value of X; thus, they are able to estimate causal effects. In the first case, one estimates an association; in the second case, one estimates an effect.

To infer a causal relationship, researchers must eliminate alternative explanations. This is an omitted variable bias problem: Researchers must account for systematic effects of potentially omitted factors that both affect the outcome and are associated with the "cause" of interest. For example, in objecting to the hypothesized causal relationship between smoking cigarettes and lung cancer, noted statistician R.A. Fisher (1935) suggested that people's genetic makeup might predispose them both to smoking and to developing lung cancer. This alternative explanation for the association between smoking and lung cancer was dismissed only after studies of identical twins revealed that a smoking twin was more likely to develop lung cancer than a nonsmoking twin (see further discussion below).

Randomization addresses the omitted variable problem by introducing a new random quantity, R, that identifies which treatment is assigned to the unit. Thus, for those units that are randomized according to the value $R = x_1$, we set $X = x_1$, and for those units that are randomized according to the value $R = x_2$, we set $X = x_2$. By introducing the new random quantity, we set the value of X and at the same time balance (on average) both observed and unobserved covariates across values of X. That is, random assignment makes treatment status independent of the other covariates, both observed and unobserved. Here, manipulation allows us to identify the direction of the causal relationship and determine whether X causes Y or Y causes X. Together, randomization and manipulation legitimize the direct causal inferences from X to Y.

This result can be formalized within the counterfactual framework described above; here, we ask what would have happened to a unit for which we set $X = x_1$ had we instead set $X = x_2$ for that unit. Again, the problem is that in the randomized experiment the unit can take only one of these values. Once we have assigned $X = x_1$, we cannot go back and investigate what would have happened had we set $X = x_2$ under the identical circumstances. Therefore, we cannot identify a causal effect for a specific unit. Note that this parallels the problem of identifying, at the individual level, the effect of racial discrimination. What we can infer, however, is the value of

$$\alpha = E(Y \mid X = x_1) - E(Y \mid X = x_2)$$

using averages across the distribution of Y. That is, we can infer the value of α, the average effect, from the difference in the expected values of Y when X is equal to x_1 and when X is equal to x_2. Randomization actually allows us to do this by using the $X = x_1$ group to measure $E(Y \mid X = x_1)$ and the $X = x_2$ group to measure $E(Y \mid X = x_2)$. That is, as alluded to above, we exploit knowledge of population averages of outcomes among multiple groups to estimate the causal effect. Because randomization has balanced the distribution of potential confounding variables across each group, this is an unbiased estimate of the average causal effect of X. Rubin (1976, 1978) provides a careful explication of this result, although different aspects of the result are implicit in the early descriptions of randomization in Neyman (1923) and Fisher (1935). As we have shown, counterfactuals go hand in hand with the notion of manipulation, but in practice they are rarely acknowledged as integral parts of the randomized controlled experiment.

Weighing Evidence from Multiple Studies

How is causality established in the absence of a perfectly designed and implemented experiment? It is possible to provide a stronger argument for causal inference by combining methods—from laboratory studies of proposed mechanisms, to field experiments demonstrating external validity, to natural experiments demonstrating policy relevance and efficacy.

Researchers can learn how the accumulation of evidence from multiple sources with a variety of research designs contributes to causal inference by examining a widely cited example of inferring causation in nonexperimental settings—the connection between smoking and lung cancer (see Box 5-2). The case of smoking and lung cancer illustrates how researchers can draw causal inferences in the absence of any single study that alone would have

BOX 5-2
Smoking and Lung Cancer

In the 1920s, physicians observed a rapid increase in death rates due to lung cancer, but it took several decades before epidemiological studies began to "confirm" what some suspected—that the rise was due to smoking. Early retrospective case-control studies that attempted to match those with and without lung cancer convinced some researchers. Others posited alternative explanations, however, and various subsequent attempts at prospective cohort studies and the expanded study of "confounder" variables did little to convince the bulk of the research community and the public more broadly that the case against cigarette smoking was closed. For example, Fisher (1935), the noted statistician and creator of experimental design, advanced the "constitutional hypothesis" of a genetic predisposition to both smoking and having lung cancer. It ultimately took a series of twin studies to set this alternative aside, but the controversy still continued.

Only after the intervention of public health experts and the ultimate downturn in lung cancer cases in the 1980s did the formal causal argument take its first form. Freedman (2000:16) argues that ultimately the strength of the case against cigarette smoking that emerged rested on "the size and coherence of the effects, the design underlying epidemiological studies, and on replications in many contexts. Great care was taken to exclude alternative explanations for the findings. Even so, the

conclusively demonstrated a causal relationship. This causal link was not accepted until findings to that effect had consistently been produced in multiple settings and in varied study designs, both observational and experimental.

Note that consistent findings across observational studies of different populations are not sufficient in and of themselves to establish a nonspurious relationship; findings must also be consistent across research designs. Otherwise the same bias, replicated across similar studies, may be responsible for an observed "effect" of the potential cause of interest. Rosenbaum (2002:224) writes: "A nontrivial replication disrupts the circumstances of the original study, to check whether the treatment produced its ostensible effect, not some irrelevant circumstance." Nontrivial replication permits researchers to exclude alternative explanations for the phenomenon of interest, and therefore to distinguish between mere associa-

argument requires a complex interplay among many lines of evidence." The coherence of the results of the numerous epidemiological studies was particularly important. There was a dose–response relationship: Persons who smoke more heavily have a greater risk of disease than those who smoke less. The risk from smoking increases with the duration of exposure. Among those who quit smoking, excess risk decreases after exposure stops.

Although one could in principle have designed an experiment in which smoking was manipulated and individuals were assigned to smoking and nonsmoking groups, this was never possible for a variety of reasons. That many of these epidemiological studies attempted to measure the causal effect in terms of odds ratios and adjusted odds ratios* was a major benefit. Despite the absence of randomized controlled experiments, the thoughtful use of controls in some studies, combined with the intervention results and the differences in the cohorts of men and women smokers, ultimately allowed for consensus on the causal conclusion (for further details, see Freedman, 2003; Gail, 1996). See also Hill (1987) for an earlier but less formal discussion of inferring causality.

* Odds are a way of expressing the probability of an event: Odds are calculated as the probability of an event divided by 1 minus the probability of that event [$\theta = \pi/(1 - \pi)$]. An odds ratio is simply a ratio of the odds of some event for each of two groups {$\omega = \theta_1/\theta_2 = [\pi_1/(1 - \pi_1)]/[\pi_2/(1 - \pi_2)] = [\pi_1 * (1 - \pi_2)]/[\pi_2 * (1 - \pi_1)]$}. We are frequently more interested in the relative risk (π_1/π_2) or the risk difference ($\pi_2 - \pi_2$). However, the odds ratio is widely used and studied.

tions and actual causal relationships. It was the consistent pattern of evidence across studies with a variety of designs and conducted in a variety of contexts that permitted researchers to conclude that the association between smoking and lung cancer is causal.

In the U.S. context, we have a history of official legalized discrimination that is not in question. Thus, we do not have far to look for causal explanations of continuing disparities between outcomes for nonwhites and whites, and many believe that simple methods produce sufficient evidence to confirm discrimination's continued existence. For example, consider a study of hiring behavior over time by an employer who, prior to the enactment of the provisions of the Civil Rights Act, had two categories of jobs—one for blacks and one for whites—with whites being paid higher wages than blacks. Following implementation of the act, the employer continued to have the same two types of jobs but now with a handful of blacks in white jobs and no whites in black jobs. Attributing these racial differences to continued discrimination may make sense intuitively. However, in the absence of explicit racial discrimination, we need more comprehensive data and powerful statistical tools to determine whether differential outcomes by race are, in fact, attributable to racial discrimination.

SUMMARY AND CONCLUSION

Racial discrimination is difficult to measure. Researchers rarely observe discriminatory behavior directly. Instead, they attempt to infer from disparate outcomes whether racial discrimination has occurred. Establishing that racial discrimination did or did not occur requires causal inference. Identifying a racial disparity and determining that an association between race and an outcome remains after accounting for plausible confounding factors is a relatively straightforward task. The real difficulty lies in going beyond the identification of an association to the attribution of cause. Insights drawn from social science theory about types of discrimination and mechanisms through which discriminatory behavior and processes may operate can play an important role here, informing research design and models and assisting researchers in identifying and testing alternative explanations (see Chapter 4). Ultimately, researchers usually must rely on the evaluation of evidence from multiple studies—considering the strength of association, consistency, and plausibility of each study's research design and findings—to draw causal conclusions.

All research methods have particular strengths and weaknesses with respect to measuring racial discrimination, particularly concerning the extent to which they support causal inferences. In particular, experimental designs facilitate causal inference but limit generalization, whereas observational designs facilitate generalization while limiting causal inference. In any research design, drawing a valid causal inference from a study requires careful specification of the assumptions and the logic underlying the inference.

In the next two chapters, we discuss in greater depth the existing literature that attempts to measure both correlations and causation between race and various outcomes. We provide a number of examples of studies that we believe to be particularly insightful or creative in the way they investigate the role of race in explaining outcomes across a variety of domains.

Conclusion: *No single approach to measuring racial discrimination allows researchers to address all the important measurement issues or to answer all the questions of interest. Consistent patterns of results across studies and different approaches tend to provide the strongest argument. Public and private agencies—including the National Science*

Foundation, the National Institutes of Health, and private foundations—and the research community should embrace a multidisciplinary, multimethod approach to the measurement of racial discrimination and seek improvements in all major methods employed.

6

Experimental Methods for Assessing Discrimination

As we discussed in Chapter 5, at the core of assessing discrimination is a causal inference problem. When racial disparities in life outcomes occur, explicit or subtle prejudice leading to discriminatory behavior and processes is a possible cause, so that the outcomes could represent, at least in part, the effect of discrimination. Accurately determining what constitutes the effect of discrimination, personal choice, and other related and unrelated factors requires the ability to draw clear causal inferences. In this chapter, we review two experimental approaches that have been used by researchers to reach causal conclusions about racial discrimination: laboratory experiments and field experiments (particularly audit studies).

OVERVIEW

Experimental Design

To permit valid causal inferences about racial discrimination, the design of an experiment and the analytic method used in conjunction with that design must address several issues. First, there are frequently intervening or confounding variables that are not of direct interest but that may affect the outcome. The effects of these variables must be accounted for in the study design and analysis. In controlled laboratory experiments, the investigator manipulates a variable of interest, randomly assigns participants to different conditions of the variable or treatments, and measures their responses to the manipulation while attempting to control for other relevant conditions or attributes. As described in the previous chapter, randomization greatly increases the likelihood of being able to infer that an observed difference between the treatment and control groups is causal. Observing a difference in outcome between the groups of participants can be the basis for a causal inference. In controlled field experiments, researchers analyze the results of a deliberately manipulated factor of interest, such as the race of an interviewer. They attempt to control carefully for any intervening or confounding variables. Random assignment of treatments to participants is frequently used to reduce any doubts about lingering effects of unobserved variables, provided, of course, that one can actually apply the randomization to the variable of interest.

In addition to the problem of credibly designing an experiment that supports a causal inference, a common weakness of experiments is a lack of external validity. That is, the results of the experiment may not generalize to individuals other than those enrolled in the experiment, or to different areas or populations with different economic or sociological environments, or to attributes that differ from those tested in the experiment.

Despite these problems, the strengths of experiments for answering some types of questions are undeniable. Even if their results may not be completely generalizable and even if they do not always capture all the relevant aspects of the issue of interest, experiments provide more credible evidence than other methods for measuring the effects of an attribute (e.g., race) in one location and on one population.

Using Experiments to Measure Racial Discrimination

Use of an experimental design to measure racial discrimination raises important questions because race cannot be directly manipulated or assigned randomly to participants. Researchers who use randomized controlled experiments to measure discrimination, therefore, can manipulate race by either varying the "apparent" race of a target person as the experimental treatment or can manipulate "apparent" discrimination by randomly assigning study participants to being treated with different degrees of discrimination.

In the first case, the experimenter varies the treatment, namely, the apparent race, by such means as by providing race-related cues on job applications (e.g., name or school attended) or by showing photographs to participants in which the only differences are skin color and facial features. The experimenter then measures whether participants respond differently under one race treatment compared with another (e.g., evaluating black versus white job applicants or associating positive or negative attributes with photographs of blacks versus whites). In such a study, the experimenter elicits responses from the participants to determine the effect of apparent race on their behavior (e.g., whether the participants engage in discriminatory behavior toward black and not white applicants). That is, they measure the behavior of potential discriminators toward targets of different races. If successful, then finding a difference in behavior would indicate an effect of race.

In the second instance, experimenters randomly assign participants to be treated differently, that is, either with or without discrimination. This type of experiment attempts to measure the response to discrimination rather than directly measure the expression of discrimination—that is, it measures the behavior of potential targets of discrimination. Because race cannot be experimentally manipulated, an explicit specification of the behavioral process is needed that allows the translation of results from such experiments into causal statements about the actual discrimination mechanism measured in the experiment (i.e., the extent to which the experimenter can manipulate some other factor related to race, such as perception). To our knowledge, no one has attempted to carry out such formal reverse reasoning, and we believe that doing so is especially crucial when arguing for the external validity of experimental results.

One of the few examples of attempts to perform similar inferential reversals is the special case of understanding odds ratios (and adjusted odds ratios) in the context of comparing retrospective and prospective studies on categorical variables. In retrospective studies, the data are collected only after the treatment has taken place, whereas in prospective studies the data are collected on possible covariates before treatment and on outcomes after the treatment. If one has both categorical explanatory and categorical response variables, one can estimate their relationship in the prospective study based on a retrospective sample. If the logistic causal model is correct, the inference about the key causal coefficient from the retrospective study is the same as if one had done a prospective sampling on the explanatory variable.[1] Those results, however, do not generalize to relationships among continuous variables.

LABORATORY EXPERIMENTS

Design

Laboratory experiments, like all experiments, include the standard features of (1) an independent variable that researchers can manipulate (i.e., assign conditions or treatments to participants); (2) random assignment to

[1]Establishing that the logistic causal model is a valid representation of the process under study is very difficult, but it is clearly necessary to draw such conclusions.

treatment conditions; and (3) control over extraneous variables that otherwise might be confounded with the independent variable of interest, potentially undermining the interpretation of causality. Laboratory experiments occur in a controlled setting, chosen for its ability to minimize confounding variables and other extraneous stimuli.

Laboratory experiments on discrimination would ideally measure reactions to the exact same person while manipulating only that person's race. As noted above, while strictly speaking one cannot manipulate the actual race of a single person, experimenters do typically either manipulate the apparent race of a target person or randomly assign subjects or study participants to the experimental condition while attempting to hold constant all other attributes of possible relevance. One common method of varying race is for experimenters to train several experimental confederates—both black and white—to interact with study participants according to a prepared script, to dress in comparable style, and to represent comparable levels of baseline physical attractiveness (see, e.g., Cook and Pelfrey, 1985; Dovidio et al., 2002; Henderson-King and Nisbett, 1996; Stephan and Stephan, 1989). Another common method of varying race involves preparing written materials and either incidentally indicating race or attaching a photograph of a black or white person to the materials (e.g., Linville and Jones, 1980).

Effects of race occur in concert with other situational or personal factors, called *moderator variables*, that may increase or decrease the effect of race on the participants' responses. In addition to manipulating a person's apparent race, for example, investigators may manipulate the person's apparent success or failure, cooperation or competition, helpfulness, friendliness, dialect, or credentials (see, e.g., Cook and Pelfrey, 1985; Dovidio et al., 2002; Henderson-King and Nisbett, 1996; Linville and Jones, 1980; Stephan and Stephan, 1989). Even more often, experimenters will manipulate features of the situation expected to moderate levels of bias toward black and white targets; examples involve anonymity, potential retaliation, norms, motivation, time pressure, and distraction (Crosby et al., 1980). Finally, the study participants frequently are black and white college students (e.g., Crosby et al., 1980; Correll et al., 2002; Judd et al., 1995).

Strengths of Laboratory Experiments

Laboratory experiments, if well designed and executed, can have high levels of internal validity for causal inference—that is, they are designed to measure exactly what causes what. The direction of causality follows from the manipulation of randomly assigned independent variables that control for two kinds of unwanted, extraneous effects: systematic (confounding) variables and random (noise) variables.

Laboratory experiments are the method of choice for isolating a single variable of interest, particularly when fine-tuned manipulation of precisely defined independent variables is required. Laboratory studies also allow precise measurement of dependent variables (such as response time or inches in seating distance). The laboratory setting gives experimenters a great degree of control over the attention of participants, potentially allowing them to maximize the impact of the manipulation in an otherwise bland environment.

Because of these fine-grained methods, laboratory experiments on discrimination are well suited to examining psychological processes. Both face-to-face interactions and processes in which single individuals react to racial stimuli are readily studied in such experiments. The most sophisticated experiments show not only the effect of some variable (e.g., expectancies) on an outcome variable (e.g., discriminatory behavior) but also the mechanism or process that mediates the effect (e.g., biased interpretations, nonverbal hostility, stereotypic associations). That is, when an experiment manipulates the apparent race of two otherwise equivalent job candidates or interaction partners (as in the interracial interaction studies described later in this chapter; see Dovidio et al., 2002; Word et al., 1974), the experiment ideally should also measure some of the proposed explanatory psychological mechanisms (such as emotional prejudices and cognitive stereotypes, either implicit or explicit), as well as the predicted discrimination (either implicit behaviors, such as nonverbal reactions, or more explicit behaviors, such as verbal reactions).

A hallmark of the better laboratory experiments is that they not only test useful theories but also show how important, compelling phenomena (e.g., the automaticity of discrimination) can and do occur. Laboratory studies often show that very small, subtle alterations in a situation can have substantial effects on important outcome variables.

Measuring Racial Discrimination

Experimenters measure varying degrees of discrimination. Laboratory measures of discrimination begin with verbal hostility (e.g., in studies of interracial aggression), which can constitute discrimination, when, for example, negative personal comments result in a hostile work environment (see Chapter 3). At the next level are disparaging written ratings of an individual member of a particular group (Talaska et al., 2003). If unjustified, such negative evaluations can constitute discrimination in a school or workplace.

At the subtle behavioral level, laboratory studies measure nonverbal indicators of hostility, such as seating distance or tone of voice (Crosby et al., 1980). Related nonverbal measures include coding of overt facial expressions, as well as measurement of minute nonvisible movements in the facial muscles that constitute the precursors of a frown. Experimenters study these nonverbal behaviors because they, too, could result in a hostile environment.

Moving up a level, laboratory measures of discriminatory avoidance include participants' choice of whether to associate or work with a member of a racial outgroup, volunteer to help an organization, or provide direct aid to an outgroup member who requests it (Talaska et al., 2003). In a laboratory setting, segregation can be measured by how people constitute small groups or choose leaders in organizational teams (Levine and Moreland, 1998; Pfeffer, 1998). Finally, aggression against outgroups can be measured in laboratory settings by competitive games or teacher–learner scenarios in which one person is allowed to punish another—an outgroup member—with low levels of shock, blasts of noise, or other aversive experiences (Crosby et al., 1980; Talaska et al., 2003).

A review of laboratory studies as of the early 1980s (Crosby et al., 1980) summarized the findings as follows. Experiments on unobtrusive forms of bias and prejudice showed that white bias was more prevalent than indicated by surveys. Experiments on helping, aggressive, and nonverbal behaviors indicated that (1) whites tended to help whites more often than they helped blacks, especially when they did not have to face the person in need of help directly; (2) under sanctioned conditions (e.g., in competitive games or administration of punishment), whites acted aggressively against blacks more than against whites but only when the consequences to the aggressor were low (under conditions of no retaliation, no censure, and anonymity); and (3) white nonverbal behavior displayed a discrepancy between verbal nondiscrimination and nonverbal hostility or discomfort, betrayed in tone of voice, seating distance, and the like. This review sparked the realization, discussed in earlier chapters, that modern forms of discrimination can be subtle, covert, and possibly unconscious, representing a new challenge to careful measurement, both inside and outside the laboratory (survey measures for these forms of discrimination are discussed in Chapter 8).

Key Examples

Since the 1980s, laboratory experiments on discrimination have concentrated more on measuring subtle forms of bias and less on examining overt behaviors, such as helping others. This shift occurred precisely because of the discrepancy between some people's overtly egalitarian responses on surveys and their discriminatory responses when they think no one is looking, or at least when they have a nonprejudiced excuse for their discriminatory behavior. In Boxes 6-1 through 6-3, we describe three of the

BOX 6-1
A Classic Laboratory Experiment on Discrimination

In a pair of experiments, Word and colleagues (1974) elicited subtle nonverbal discriminatory behaviors from white interviewers against black job applicants and then demonstrated that such behaviors used against white applicants elicited behaviors stereotypically associated with blacks. The researchers first asked white college students to interview black and white high school applicants for a team that would plan a marketing campaign. Interviewers expected to see several applicants; the first applicant always was white, followed by black and white applicants in a randomly counterbalanced order. Unbeknownst to the interviewer participants, the applicants were confederates of the experimenters, trained to respond in a standard way. Debriefing indicated that study participants were unaware of its purposes or that the alleged applicants were confederates (probably aided by the sequence including two white applicants and one black applicant). Extensive debriefing indicated no suspicion about the confederates.

The interviewers' nonverbal behavior indicated less immediacy (i.e., greater discomfort and less warmth) toward black than white applicants on a number of measures scored by judges behind one-way mirrors: greater physical seating distance, shorter interviews, and more speech errors. Although judges were not blind to the race of the confederates and therefore may have been influenced in their coding of the white interviewers' behavior, three points suggest that the researchers were able to

best examples of controlled laboratory experiments on discrimination, ranging from simpler classic to more recent sophisticated studies. In a classic example, Word et al. (1974) created working definitions of race and discrimination to investigate subtle yet potentially powerful effects of stereotypical expectations hypothesized to result in discrimination (see Box 6-1). Another famous experiment showed that researchers can study social perception processes hypothesized to underlie discrimination, in which people see what they want to see by interpreting ambiguous evidence to fit their stereotypical biases (Darley and Gross, 1983; see Box 6-2). And in a final experiment, Dovidio et al. (2002) showed that implicit forms of prejudice tended to lead to implicit but potentially important forms of discrimination, whereas explicit forms of prejudice tended to lead to explicit forms of discrimination (see Box 6-3).

obtain fairly unbiased coding: (1) the coding consisted largely of physical measurement (e.g., seating distance, number of minutes) and counting (e.g., number of speech errors); (2) judges were unaware of the study's hypotheses; and (3) replications of the coded behaviors produced the expected results in a second study.

In the second experiment, white interviewers were confederates trained to behave nonverbally in either a more or less immediate way toward naive white applicants; that is, they were trained to treat some of the white applicants as the black applicants in the previous study had been treated. White applicants treated as if they were black reciprocated with greater seating distance and more speech errors. They perceived the interviewer to be less friendly and less adequate. They also performed worse in the interview; were judged less adequate for the job; and appeared less calm, composed, and relaxed.

The overall point of this pair of experiments—the basic methods of which have since been replicated repeatedly—is that researchers can investigate how simulating discrimination against whites can bring about the very behaviors that are stereotypically associated with blacks (or another disadvantaged racial group), and they can measure the hypothesized mechanisms involved—that is, subtle nonverbal cues unlikely to be analyzed consciously by either perceiver or target. Moreover, researchers can mimic the employment interview context to examine the potentially large effects that nonverbal forms of discrimination are hypothesized to have on people's ability to obtain a job.

Other provocative recent experiments have shown that actual discriminatory behavior can follow from subliminal exposure to racial and other demographic stimuli (Bargh et al., 1996). This work has revealed that exposure to concepts and stereotypes at speeds too fast for conscious recognition primes relevant behavior, even though participants cannot remember or report having seen the priming stimuli. For example, researchers randomly assigned participants to see, at subliminal speeds, words related to rudeness or neutral topics and showed that those participants exposed to rude words responded more rudely to an experimenter. In a parallel experiment, subliminal exposure to photographs of unfamiliar black male faces, as compared with white ones, was followed by more rude, hostile behavior when the white experimenter subsequently made an annoying request. Similar results have been demonstrated for exposure to phenomena related to being

BOX 6-2
Perceptions of Academic Performance

In a laboratory experiment conducted by Darley and Gross (1983), participants viewed a child depicted in a 6-minute videotape as coming from either a high or low socioeconomic background, based on the setting in which she was shown playing. When asked to rate her academic performance, they acknowledged not having enough information and rated her ability at grade level. Other participants saw the initial 6-minute videotape depicting socioeconomic status but also saw an additional 12-minute videotape that depicted the child taking an oral test on which her performance was mixed. Participants shown the second video after the first no longer demurred regarding the child's academic performance. Instead, they rated her performance as well below grade level if they had viewed the 6-minute video depicting low socioeconomic status and at grade level if they had seen the video depicting high socioeconomic status. Control participants shown the test video alone, and not also the socioeconomic status video, rated the child's performance at about grade level.

Thus, the researchers were able to show how people perceived the academic performance tape (itself quite neutral) through the lens of their expectations, convincing themselves that they had evidence on which to base their biased judgments. Although the child on the tape was white, the applicability of this sort of socioeconomic status-based stereotype to racially tainted judgments of academic performance appears clear, and manipulating such variables sheds light on hypothesized processes of discrimination. Methods for assessing this kind of perceptual confirmation process have been replicated repeatedly. For example, in a study conducted by Sagar and Schofield (1980), black and white sixth-grade boys viewed depictions of various ambiguously aggressive behaviors by black and white actors. Participants read identical verbal descriptions of four ambiguously aggressive incidents common in middle schools: bumping in the hallway, requesting another student's food, poking in the classroom, and using another's pencil without permission. The race of actors and targets was not specified verbally, but each incident was accompanied by one of four drawings of the event, identical except for the depicted race of the actor and target. Participants saw each incident only once and each in just one of the four possible combinations of actor race and target race. Participants rated how mean, threatening, friendly, and playful each incident was. Researchers were able to show that all the participants, regardless of race, rated the behaviors as more mean and threatening when a black child enacted them than when a white child did.

elderly, which resulted in participants walking more slowly to the elevator after the experiment. The point is that researchers can manipulate racial cues without participants' conscious awareness and measure subtle forms of behavior that, if occurring selectively toward members of one racial group or another, could constitute a hostile environment form of discrimination. Other more direct forms of discrimination are also possible to measure in such experiments, such as making negative comments in a job interview.

These examples illustrate the range of aspects of racial discrimination that can be examined in laboratory settings. Such experiments can manipulate racial and moderator variables; test various hypothesized mechanisms of discrimination, such as attitudes; and assess various hypothesized manifestations of discrimination, including verbal, nonverbal, and affiliative responses. They can also simulate pieces of real-world situations of interest, such as job applications and others. Most of the phenomena studied in experiments on race discrimination have been replicated in studies of gender discrimination and sometimes age, disability, class, or other ingroup–outgroup variations. Research indicates that gender, race, and age are the most salient, immediately encoded social categories (Fiske, 1998).

Limitations of Laboratory Experiments

Laboratory experiments usually are limited in time and measurement, so they generally do not aim to answer questions about behavior over long periods of time or behavior related to entire batteries of measures. The purpose of a laboratory experiment may include one or more of the following: (1) to demonstrate that an effect indeed can occur, at least under some conditions, with some people, for some period of time; (2) to create a simulation or microcosm that includes the most important factors; (3) to create a realistic psychological situation that is intrinsically compelling; or (4) to test a theory that has obvious larger importance.

Laboratory experiments are also at risk for various biases related to the settings in which they occur. For example, they may be set up in such a narrow, constraining way that the participants have no choice but to respond as the experimenters expect (Orne, 1962). Crafting more subtle manipulations and providing true choice in response options can sometimes be used to limit the potential biases in such cases. In addition, the experimenter may inadvertently bias presentation of the manipulations and measures, so that participants are equally inadvertently induced to confirm the hypotheses (Rosenthal, 1976). This problem can often be addressed using double-blind methods, in which experimenters as well as participants are not aware of the treatment assigned to them. Participants may also worry

BOX 6-3
The Effect of Psychological Mechanisms on Measures of Discriminating Behavior

Laboratory experiments can create working definitions of manipulated race, randomly assign participants to interact with black or white confederates, and measure a variety of proposed psychological mechanisms (implicit and explicit attitudes) to determine their effect on various types of discriminatory behavior. For example, Dovidio et al. (2002) conducted a multiphase experiment on how whites' explicit and implicit racial attitudes predict bias and perceptions of bias in interracial interactions. At the beginning of the term, white college students completed a 20-item standardized measure of prejudice, the Attitudes Toward Blacks Scale. Later in the semester, 40 students (15 male and 25 female) participated in what they believed to be two separate studies. In the first, a decision task required participants to respond as quickly as possible—after the letter P or H was displayed on a computer screen—as to whether a given word displayed for each trial could ever describe a person or a house. Unbeknownst to them, on critical trials versus practice trials the letter P was preceded by a standardized schematic sketch of a black or white man or woman, presented at subliminal speeds (0.250 seconds). This level of presentation has been shown repeatedly to prime relevant associations in memory and, in particular, stereotypes. As in countless other studies (e.g., see Fazio and Olson, 2003), the findings in this study revealed subtle forms of stereotypic association when people responded more quickly to negative words ("bad," "cruel," "untrustworthy") preceded by a black face and to positive words ("good," "kind," "trustworthy") preceded by a white face, and more slowly to the converse combinations. As is typical with this method, no participant reported being aware of the subliminal faces. Such studies show how researchers can measure automatic and unconscious racial bias, regardless of expressed levels of prejudice (Devine, 1989). At this point, then, the experimenters had access to two kinds of attitudes—the explicit ones expressed on the questionnaire and the implicit ones suggested by the participants' speed of stereotypic associations. These are the psychological causes of different kinds of discrimination hypothesized in the next step.

In what participants assumed to be a separate study focused on acquaintance processes, the participants met separately with two interaction partners—one white and one black—for a 3-minute conversation about dating in the current era. Five white and four black student confederates, trained to behave comparably to each other, played the role of interaction partners. All were unaware of the study's hypotheses and the participants' levels of implicit and explicit prejudice. After each interaction, both the participant and the confederate (in separate rooms) completed scales assessing their own and each other's perceived friendliness (pleasant and not cold, unfriendly, unlikable, or cruel). Two coders used the same scales to rate, separately, participants' verbal and nonverbal behavior, respectively, from audiotapes and from videotapes on which only the participant was visible. Two more coders rated participants' overall friendliness from audio and video information combined. Analyses compared the differences in the participants' responses to the white and black confederates as rated by the participants themselves, the confederates, and the observers.

Two patterns of response emerged: one an explicit and overt sequence of processes, and the other an implicit and subtle sequence of processes. The explicit sequence involved overt measures of verbal behavior. White participants' scores on the attitudes questionnaire and their self-reported friendliness (both measures of explicit, overt prejudice) correlated with each other; that is, whites' self-reported attitudes predicted bias in verbal friendliness toward black relative to white confederates. These measures also correlated with verbal friendliness as rated by observers from audiotapes (a measure of explicit, overt discriminatory behavior).

In contrast, the implicit sequence of processes was indicated by responses to subliminal primes (an implicit, subtle measure of prejudice), which correlated significantly with a series of implicit, subtle forms of discriminatory behavior: nonverbal behavior rated by observers from silent videotapes, confederate perceptions of participants' friendliness, and overall friendliness rated by other observers, which also correlated significantly with each other. In other words, whites' implicit attitudes predicted their bias and others' perceptions of bias in nonverbal friendliness. None of the explicit and implicit measures correlated significantly with each other, indicating that the implicit and explicit sequences are independent. Each sequence is important: Effect sizes were moderate to large by social science standards.

about whether their behavior is socially acceptable (Marlow and Crowne, 1961) and fail to react spontaneously. Nonreactive, unobtrusive, disguised measurement can avert this problem. It is worth noting that not all of these issues are unique to the laboratory. Many of the potential biases and artifacts of laboratory experiments also occur at least as often in other kinds of experiments (e.g., field experiments, which we turn to next), as well as with nonexperimental methods (natural experiments and observational studies, such as surveys).

Translating Experimental Effects

Laboratory experiments are useful for measuring psychological mechanisms that lead to discriminatory behavior (e.g., implicit or explicit stereotypes), but they do not describe the frequency of occurrence of such behavior in the world. They cannot, by their nature, say how often or how much a particular phenomenon occurs, such as what proportion of a racial disparity is a function of discriminatory behavior. Thus, they can be legitimately criticized on the grounds of low external validity—that is, limited generalizability to other samples, other settings, and other measures. Laboratory experimenters can sometimes make a plausible case for generalizability by varying plausible factors that might limit the applicability of the experiment. For example, if there are theoretically or practically compelling reasons for suspecting that an effect is limited to college sophomores, one might also replicate the study with business executives on campus for a seminar or retirees passing through for an Airstream conference. But laboratory experiments rarely randomly sample participants from the population of interest. Thus by themselves they cannot address external validity, and it is an empirical question whether or how well their findings translate into discrimination occurring in the larger population. In well-designed and well-executed experiments, the effects of confounding variables are randomized, allowing researchers to dismiss competing explanations as unlikely, but they are not entirely eliminated. For this reason, replication is important. In the study of discrimination, there are many laboratory experiment results that do not generalize in field settings. Findings either may diminish or not hold up over time. However, many other effects tested both in the laboratory and in the field have been consistent, some showing even stronger effects in the field (Brewer and Brown, 1998; Crosby et al., 1980; Johnson and Stafford, 1998).

FIELD EXPERIMENTS

Design

Field experiments have many of the standard features commonly found in laboratory experiments. The term *field experiment* refers to any fully randomized research design in which people or other observational units found in a natural setting are assigned to treatment and control conditions. The typical field experiment uses a two-group, post-test-only control group design (Campbell and Stanley, 1963). In such a design, people are randomly assigned to treatment and control groups. An experimental manipulation is administered to the treatment group, and an outcome measure is obtained for both treatment and control groups. Because of random assignment, differences between the two groups provide some evidence of an effect of the manipulation. However, because no preexperiment measure for the outcome is obtained (which is an option in laboratory experiments), one cannot be altogether sure whether the groups are similar prior to the experiment. Nonetheless, randomization protects against this problem because it ensures that, *on average*, the two groups are similar except for the treatment.

Field experiments are attractive and often persuasive because, when done well, they can eliminate many of the obstacles to valid statistical inference. They can measure the impact of differential treatment more cleanly than nonexperimental approaches, yet they have the advantage of occurring in a realistic setting and hence are more directly generalizable than laboratory experiments. Furthermore, for measuring discrimination, they appear to reflect the broader public vision of what discrimination means—the treatment of two (nearly) identical people differently.

The social scientific knowledge necessary to design effective field experiments is stronger in some areas than in others. For example, our knowledge of the mechanisms and incentives underlying real estate markets is arguably more advanced than our knowledge of the incentives underlying labor markets (Yinger, 1995). Hence, our ability to use field experiments is correspondingly stronger for measuring behavior in housing markets than in other areas. We therefore focus our discussion below on a common methodology—audit or paired testing—used particularly to assess discrimination in housing markets as well as in other areas. With the exception of a study we describe later (in Box 6-5), we do not review other types of field experiments in the domain of racial discrimination.

Audit or Paired-Testing Methodology[2,3]

Audit or paired-testing methodology is commonly used to measure the level or frequency of discrimination in particular markets, usually in the labor market or in housing (Ross, 2002; for a summary of paired-testing studies in the labor and housing markets, see Bendick et al., 1994; Fix et al., 1993; Neumark, 1996; Riach and Rich, 2002). Auditors or testers are randomly assigned to pairs (one of each race) and matched on equivalent characteristics (e.g., socioeconomic status), credentials (e.g., education), tastes, and market needs. Members of each pair are typically trained to act in a similar fashion and are equipped with identical supporting documents. To avoid research subjects becoming suspicious when they confront duplicate sets of supporting documents, researchers sometimes vary the documents while keeping them similar enough that the two testers have equivalent levels of support.

As part of the study, testers are sent sequentially to a series of relevant locations to obtain goods or services or to apply for employment, housing, or college admission (Dion, 2001; Esmail and Everington, 1993; Fix et al., 1993; National Research Council, 1989; Schuman et al., 1983; Turner et al., 1991a, 1991b; Yinger, 1995). The order of arrival at the location is randomly assigned. For example, in a study of hiring, testers have identical résumés and apply for jobs, whereas in a study of rental housing, they have identical rental histories and apply for housing. Once the study has been completed, researchers use the differences in treatment experienced by the testers as an estimate of discrimination.

To the extent that testers are matched on a relevant set of nonracial characteristics, systematic differences by the race of the testers can be used to measure discrimination on the basis of race. *Propensity score matching* is sometimes used when there are too many relevant characteristics on which to match on every one. In propensity score matching, an index of similarity is created by fitting a logistic regression with the outcome variable being race and the explanatory variables being the relevant characteristics on which one wishes to match. Subjects of one race are then paired or matched with subjects of the other race having similar fitted logit values—the pro-

[2]In the following discussion on audit studies, we draw heavily on a commissioned paper by Ross and Yinger (2002) examining the challenges involved in measuring discrimination for both scholars and enforcement officials.

[3]The term *audit* is used in a research context to refer to direct evidence of discrimination in a particular market (see Fix et al., 1993, for an overview of auditing). The term *paired testing* is used to refer to studies of discrimination conducted in an enforcement context to monitor civil rights compliance. Matching or pairing is also used more generally to refer to the widely used statistical method of comparing outcomes from individuals or groups of individuals that are similar in attributes other than the one of interest (e.g., race).

pensity score index (see Rosenbaum, 2002, and the references therein for a more complete description).

Paired-testing studies use an experimental design in natural settings to obtain information on apparently real outcomes and to assess the occurrence and prevalence of discrimination. An advantage to using paired tests is that individuals are matched on observed characteristics relevant to a particular market. Effective matching decreases the likelihood that differences are due to chance rather than discrimination because many factors are controlled for.

Paired testing is used in audit studies, such as the U.S. Department of Housing and Urban Development's (HUD's) national study of housing discrimination, to estimate overall levels of discrimination against racial and ethnic minorities. Audit studies can be highly effective enforcement tools for assessing treatment or detecting unfavorable treatment of members of disadvantaged groups (see Ross and Yinger, 2002).[4] Studies in the housing market (e.g., Wienk et al., 1979; Yinger, 1995) and in the labor market (e.g., Bendick et al., 1994; Cross et al., 1990; Neumark, 1996; Turner et al., 1991b) using the paired-testing methodology provide evidence of discrimination against racial minorities (see National Research Council, 2002b; Ross and Yinger, 2002). In the case of housing, these studies might involve selecting a random sample of newspaper advertisements and then investigating the behavior of real estate agencies associated with these advertisements (Ross and Yinger, 2002). Employment audits are similarly based on a random sample of advertised jobs. While providing the generality valued by researchers, these studies also make it possible to observe the behavior of individual agencies or firms. This approach has been applied to other areas as well (see the examples in the next section).

Key Examples

Much of the use of audit or paired-testing methodology to study discrimination flows primarily from federal investigations concerning housing discrimination. National results of the 2000 Housing Discrimination Study (2000 HDS), conducted by the Urban Institute for HUD, show that housing discrimination persists, although its incidence has declined since 1989 for African Americans and Hispanics. Non-Hispanic whites are consistently favored over African Americans and Hispanics in metropolitan rental and

[4]Ross and Yinger (2002) posit that because only a single audit is typically conducted for a given firm, audit studies can pose challenges for enforcement officials. One solution is to combine results from an audit study based on a random sample with results from audits of additional firms found to discriminate, thereby reducing the enforcement burden on targets of discrimination who file specific complaints.

sales markets (Turner et al., 2002b); similarly, Asians and Pacific Islanders in metropolitan areas nationwide (particularly homebuyers) face significant levels of discrimination (Turner et al., 2003; see Box 6-4 for a brief history of housing audits). In another example, Yinger (1986) studied the Boston housing rental and sales markets in 1981. In the rental market, whites discussed 17 percent more units with a rental agent and were invited to inspect 57 percent more units than blacks. In the sales market, whites discussed 35 percent more houses and were invited to inspect 34 percent more houses; moreover, the difference in treatment was larger for low-income families and families with children. Yinger also found substantial variation in treat-

BOX 6-4
Housing Audits

Perhaps the most common method of assessing discrimination in housing is the fair housing audit. This approach, also referred to as paired testing in an enforcement context, is used in fair housing enforcement by private fair housing groups, public fair housing agencies, and the U.S. Department of Justice (Yinger, 1995). HUD has conducted several times what is by far the largest field experiment using matched-pair methodology—the Housing Discrimination Study (HDS). Results of the most recent 2000 HDS (released in November 2001) show that housing discrimination has declined since 1989 for African Americans and Hispanics, but it nonetheless persists: Non-Hispanic whites are consistently favored over African Americans and Hispanics in metropolitan rental and sales markets (Turner et al., 2002b). Similarly, Asians and Pacific Islanders in metropolitan areas nationwide (particularly homebuyers) face significant levels of discrimination (Turner et al., 2003; also, see National Research Council, 2002b, for a review of the 2000 HDS design).

Housing audits conducted after the passage of the Fair Housing Act (Title VIII of the Civil Rights Act of 1968) have been used to address discrimination and ensure equal opportunity in housing. The first audits were carried out by local fair housing organizations, often for purposes of enforcement but also to gather information. Results of the earliest audits were impaired by small sample sizes, nonrandom assignment methods, and failure to use standardized instruments and procedures. However, practices and methods gradually improved, and the cumulative body of work consistently showed that African Americans continued to suffer from various forms of housing discrimination despite the legal prohibition of such discrimination (see Galster, 1990a, 1990b, for reviews of local studies).

ment across neighborhoods. Taken together, these results document significant discrimination in the housing market.

As reported by Ross and Yinger (2002) and by Riach and Rich (2002), although the typical audit study concerns housing (e.g., Donnerstein et al., 1975; Schafer, 1979; Wienk et al., 1979; Yinger 1986), researchers have used variants of the design described above to examine discrimination in other areas. Areas studied include the labor market (Turner et al., 1991b), entry-level hiring (Cross et al., 1990), automobile purchases (e.g., Ayres and Siegelman, 1995), helping behaviors (Benson et al., 1976), small favors (Gaertner and Bickman, 1971), being reported for shoplifting (Dertke et al.,

The first attempt to measure housing discrimination nationally was carried out by HUD in the HDS of 1977. This study covered 40 metropolitan areas chosen to represent areas with central cities that were at least 11 percent black. The study confirmed the results of earlier local housing audits and demonstrated that discrimination was not confined to a few isolated cases (Wienk et al., 1979).

The 1977 HDS was replicated in 1988. Twenty audit sites were randomly selected from metropolitan areas having central-city populations exceeding 100,000 and that were more than 12 percent black. Real estate ads in major metropolitan newspapers were randomly sampled, and realtors were approached by auditors who inquired about the availability of the advertised unit and other units that might be on the market. The study covered both housing rentals and sales, and the auditors were assigned incomes and family characteristics appropriate to the housing unit advertised (Turner et al., 1991a).

The resulting data offered little evidence that discrimination against blacks had declined since the 1977 assessment (Yinger, 1993). The incidence of discriminatory treatment (defined as the percentage of encounters in which discrimination occurred) was over 50 percent in both the rental and the sales markets. The severity of the discrimination was also very high (severity being the number of units made available to whites but not blacks). Across indicators (e.g., number of advertised units shown, number of other units mentioned or shown, and location of units shown), between 60 and 90 percent of the housing units made available to whites were not brought to the attention of blacks. Over the course of the 1990s, various researchers carried out housing audits in different metropolitan areas using various methods (Galster, 1998; Massey and Lundy, 2001; Ondrich et al., 2000).

1974), obtaining a taxicab (Ridley et al., 1989), preapplication behavior by lenders (Smith and Delair, 1999; Turner et al., 2002a), and home insurance (Squires and Velez, 1988; Wissoker et al., 1997).

In an example involving automobile purchases, Ayres and Siegelman (1995) sent 38 testers (19 pairs) to 153 randomly selected Chicago-area new-car dealers to bargain over nine car models. Testers bargained for the same model (a model of their mutual choice) at the same location within a few days of each other. In contrast with the common paired-testing design, pair membership was not limited to a single pair; instead, testers were assigned to multiple pairs. Also, testers did not know that the study was intended to investigate discrimination or that another tester would be sent to the same dealership. Testers were randomly allocated to dealerships, and the order of their visits was also randomly assigned. The testers were trained to follow a bargaining script in which they informed the dealer early on that they would not need financing. They followed two different bargaining strategies: one that depended on the behavior of the seller and another that was independent of seller behavior.

Ayres and Siegelman found that initial offers to white males were approximately $1,000 over dealer cost, whereas initial offers to black males were approximately $1,935 over dealer cost. White and black females received initial offers that were $1,110 and $1,320 above dealer cost, respectively. Final offers were lower, as expected, but the gaps remained largely unchanged. Compared with white males, black males were asked to pay $1,100 more to purchase a car, black females were asked to pay $410 more, and white females were asked to pay $92 more. These examples of evidence gleaned on market discrimination show the value of paired-testing methods for studying discrimination.

In Box 6-5, we provide an example of a field experiment on job hiring (Bertrand and Mullainathan, 2002) that emulates some of the best features of laboratory and audit studies. This study uses a large sample and avoids many of the problems of audit studies (e.g., auditor heterogeneity) by randomly assigning race to different résumés. It is a particularly good example of the possibilities of field study methodology to investigate racial discrimination.

Limitations of Audit Studies

Ross and Yinger (2002) discuss two main issues raised by researchers concerning the use of paired-testing methodology. They are (1) the accuracy of audit evidence and (2) its validity, particularly with respect to the target population. It is also worth noting that such studies typically require extensive effort to prepare and implement. They can be very expensive.

The Accuracy Issue

Many claim that the designs of audit studies are not true between-subjects experiments because research subjects (e.g., employer or housing agent) are not assigned to treatment or control groups but are exposed to both treatment and control (see Chapter 7 for a discussion of issues in repeated-measures designs). Also, although the order of exposure for each subject is randomized so that it should balance out, the time lapse between exposures makes it possible for the difference to be unrelated to the concept of focus (i.e., discrimination). In the time between two visits to an establishment, for example, someone else other than a tester may take the job or apartment of interest.

In the housing market, newspaper advertisements are used as a sampling frame (National Research Council, 2002b), but they may not accurately represent the sample of houses that are available or affordable to members of disadvantaged racial groups. Newspaper advertisements can be limiting because the sampling frame is restricted to members of disadvantaged racial groups who respond to typical advertisements and are qualified for the advertised housing unit or job. This limited sample may lead to a very specific interpretation of discrimination. For example, members of the sample may not be aware of alternative search strategies or know of other available housing units or jobs of interest. The practical difficulties associated with any sampling frame other than newspaper advertisements (and the associated steps of training auditors and assigning characteristics to them) are difficult to overcome.

The Validity Issue

Inferential target: estimating an effect of discrimination. Researchers have also debated the validity of audit studies (see the discussion in Ross and Yinger, 2002). Heckman and colleagues criticize the calculation of measures of discrimination (Heckman, 1998; Heckman and Siegelman, 1993). They argue that an estimate of discrimination at a randomly selected firm (or in an advertisement) does not measure the impact of discrimination in a market. Rather, discrimination should be measured by looking at (1) the average difference in the treatment of disadvantaged racial groups and whites or (2) the actual experience of the average member of a disadvantaged racial group, as opposed to examining the average experience of members of disadvantaged racial groups in a random sample of firms (i.e., the focus should be on the average across the population of applicants rather than the population of firms). Both of these proposed approaches to measuring discrimination are valid, but each has limitations.

Researchers typically determine the incidence of discrimination by mea-

BOX 6-5
Combining Features of Laboratory and Audit Studies

Bertrand and Mullainathan (2002) conducted a large-scale field experiment on job hiring by sending résumés in response to over 1,300 help-wanted advertisements in Boston and Chicago newspapers (submitting four résumés per ad). In all they submitted 4,890 résumés. For each city, the authors took résumés of actual job seekers, made them anonymous, and divided them into two pools based on job qualifications—high and low. Two résumés from each pool were assigned to each advertisement, and race was randomly assigned within each pair. Thus, they randomly assigned white-sounding names (e.g., Allison and Brad) to two of the résumés and black-sounding names (e.g., Ebony and Darnell) to the remaining two résumés. This crucial randomization step breaks the tie between the résumé characteristics and race. Addresses were also randomized across résumés so that the ties between race and neighborhood characteristics and résumé attributes and neighborhood characteristics were also broken. Thus for each ad the researchers were able to observe differential callbacks by race both within and between the high- and low-qualified résumé pools.

Using callback rate as the outcome of interest, the authors found that on average, applicants with white-sounding names received 50 percent more callbacks than applicants with black-sounding names. Specifically, the researchers found a 12 percent callback rate for interviews for "white" applicants compared with a 7 percent callback rate for interviews for "black" applicants. They also found that higher-quality résumés yielded significant returns for white applicants (14 percent callback rate for white applicant/high-quality résumés versus 10 percent for white applicant/low-quality résumés) but not for black applicants (7.7 percent callback rate for black applicant/high-quality résumés versus 7.0 percent for black applicant/low-quality résumés). The authors concluded that for blacks having more productive skills may not necessarily reduce discrimination.

By randomizing the assignment of race, the authors made it possible to directly estimate the usual missing counterfactual—whether a callback would have been received if the résumé had belonged to an applicant likely to be perceived as being of the other race. Two résumés were selected from each pool (high- and low-qualified) because the same résumé could not be sent in response to a single advertisement with different names and addresses attached but otherwise identical content. Because race was randomized within each quality pair, any difference by race in the résumé quality (within a quality pool) for a particular advertisement could be expected to average out over a large number of advertisements. Thus the outcomes of the two résumés within a quality level could be compared, and the average of these comparisons could provide an estimate of the effect of race on callbacks within each quality level, which would also provide an estimate of the effect of any interaction between race and qualifications.

More formally, with the analysis done at the résumé level, the causal effect of interest is as follows, where *CB* stands for callback, *W* for white, and *B* for black:

$$E[CB\,(\text{Race} = W)] - E[CB\,(\text{Race} = B)].$$

Because race was randomized within quality levels, which were assigned to particular advertisements within particular cities, this causal difference by race can be estimated within each of those categories by calculating

$$(1/n)\sum_{city}\sum_{ads}\sum_{quality}(CB_W - CB_B) = (1/n_W)\sum_{city}\sum_{ads}\sum_{quality}CB_W - (1/n_B)\sum_{city}\sum_{ads}\sum_{quality}CB_B.$$

In addition, estimates for subpopulations within a quality level or city or type of advertisement can be estimated by summing just over those subpopulations.

These observations about the design and estimand of interest, along with the assumption of unit treatment additivity for city, advertisement, and a quality-by-race interaction effect, suggest the following model:

$$\Pr(\text{Callback}_{ijkl}) = f(\text{City}_i + \text{Quality}_j + \text{Race}_k + \text{Quality}_j \times \text{Race}_k),$$

where $f(.)$ is a function that produces a probability of callback. The outcome is measured with errors ε_{ijkl} that are correlated within an advertisement, as they would be for observations within a cluster in a sample. Alternatively, the advertisements themselves can be included in the model, which makes the error terms independent. This model would take the form

$$\Pr(\text{Callback}_{m(i)jkl}) = f(\text{City}_i + \text{Ad}_{m(i)} + \text{Quality}_j + \text{Race}_k + \text{Quality}_j \times \text{Race}_k),$$

where the extra subscript on the Ad variable acknowledges the fact that advertisements are nested within a city.

This design has several advantages over audit studies. One advantage is the ability to use a large number of résumés, as opposed to a smaller number of auditors, and thus the ability to send those résumés out to a large number of employers. The most significant advantage of this design is the ability of the researchers to randomize race, or a proxy for race, instead of trying to match actual people on as many characteristics as possible. The significant constraint this strategy imposes is that the outcome measured—receiving a callback from an employer—is from the early stages of the job search, as is necessary when the only contact is a résumé.

continued

BOX 6-5 Continued

One concern regarding this study is that there may be real or perceived characteristics, such as class, that are associated with distinctively African American or distinctively white names that differ from the real or perceived characteristics of these groups more generally. The authors checked whether differences in mothers' educational status by particular distinctive names correlated with differences in callback rates for particular names and found no significant correlation. However, this check does not address the present concern; rather, it suggests that the researchers have the data to determine whether the educational status of mothers who give their children distinctively African American names differs from that of both African American and white mothers who do not give such names. The authors also report having conducted a survey in Chicago in which respondents were given a name and asked to assess features of the person. This was done to check that respondents identified the correct race with the racially distinctive name, but also could have been used to check whether there are perceptions of other characteristics that vary within race based on how racially distinct a name is.

suring (1) the proportion of cases in which a white tester reports more favorable treatment than a nonwhite tester reports (gross adverse treatment) or (2) the difference between the proportion of cases in which a white tester reports favorable treatment and the proportion of cases in which a nonwhite tester reports favorable treatment (net adverse treatment) (for further discussion of these measures, see Fix et al., 1993; Heckman and Siegelman, 1993; Ondrich et al., 2000; Ross, 2002). Because statistical measures are "model-based" aggregates, net measures correctly measure the parameters in those models conditional on important stratifying variables. The gross measure may provide useful supplemental information to the net measure if the balancing disparities are large.

Ross and Yinger (2002) note that it would be valuable to know the true experiences of members of disadvantaged racial groups on average, but such information could not reveal the extent to which these individuals change their behavior to avoid experiencing discrimination. As a result, discrimination encountered by averaging over members of a disadvantaged racial group is not a complete measure of the impact of racial discrimination (Holzer and Ludwig, 2003). It is valuable to determine how much discrimination exists before such behavioral responses take place—which is the amount estimated using paired testing—and whether discrimination arises under certain circumstances.

The key observation of Murphy (2002) relates to the inferential target: Are we interested in estimating an overall or a market-level discrimination effect? Several distinct effects might be estimated, and they need to be distinguished because the estimates that result will not necessarily be identical. What is the appropriate population of real estate agents or ads from which to sample? Do we want to use only those agents that minorities actually visit? If past discrimination affects choice of agent, this population may vary from the population of agents selling houses that members of a nonwhite population could reasonably afford. Thus, the estimated effect of discrimination will be different under these alternative sampling strategies. Would it make sense to sample from agents or ads that could not reasonably be expected to be appropriate for most members of the nonwhite population? Murphy recommends ascertaining "discrimination in situations in which Blacks are qualified buyers" (2002:72).

Auditor heterogeneity. Heckman and colleagues (Heckman, 1998; Heckman and Siegelman, 1993) also argue that average differences in treatment by race may be driven by differences in the unobserved characteristics of testers (i.e., auditor heterogeneity) rather than by discrimination.[5] Such characteristics (e.g., accent, height, body language, or physical attractiveness) of one or the other member of the pair may have a significant impact on interpersonal interactions and judgments and thus lead to invalid results (Smith, 2002). The role of these characteristics cannot be eliminated because of the paucity of observations of the research subjects. Ross (2002) addresses the problem by suggesting that, instead of trying to match testers exactly (which is virtually impossible), one can train testers to ensure that their true characteristics, as opposed to their assigned characteristics, have little influence on their behavior during the test.

Murphy (2002) addresses most of the issues raised by Heckman (1998) and discussed above. She lays out a framework showing that "as long as audit pairs are matched on all qualifications that vary in distribution by race, audit results averaged over realtors, circumstances of the visits, and auditors can be viewed as an unbiased estimate of overall-level discrimination" (Murphy, 2002:69). Murphy formally delineates the circumstances under which an estimate of discrimination will be erroneous if the researcher fails to account for individual auditor characteristics that do not vary in distribution by race and therefore were not used in the matching process.

The problem is the effect of the heterogeneity among applicants and agents. The strategy of matching on *all* characteristics that vary in distribution by race—including observed, unobserved, and unobservable character-

[5]Note that such characteristics could also lead to an understatement of discrimination.

istics—substitutes for randomization. The problem, of course, is that we do not know whether we have in fact matched on all characteristics that vary by race. If all unmatched characteristics have the same distribution across racial groups, and if the auditors were selected to be representative of the distribution of these characteristics, we will have managed to balance the covariates across racial groups and can estimate an unbiased effect of race. But as Heckman and others note, there are a variety of reasons to believe that this goal of matching is elusive.

Heckman and Seigelman (1993) make the point that the problem of auditor heterogeneity poses a challenge particularly for employment audits, as well as for studies of wage discrimination, because the determinants of productivity within a firm are not well understood and are difficult to measure. Ross and Yinger (2002:45) note: "Heckman and Siegelman argue that matching may ultimately exacerbate the biases caused by unobserved auditor characteristics because those characteristics are the only ones on which [testers] differ; however, the direction and magnitude of this type of bias [are] not known." Heckman and his colleague further argue that the factors that employers use to differentiate applicants are not well known; thus, equating testers on those factors can be difficult, if not impossible. This lack of knowledge may make experimental designs particularly problematic for labor market behaviors. However, it does not affect designs in areas with a well-known or identifiable set of legitimate cues to which establishments or authorities may respond (e.g., the rental market).

There are several other problems associated with paired testing. First, paired testing cannot be used to measure discrimination at points beyond the entry level of the housing or labor market. Examples are job assignments, promotions, discharges, or terms of housing agreements and loans. Second, the assignments and training provided to testers may not correspond to qualifications and behaviors of members of racially disadvantaged groups during actual transactions. Third, actual home or job seekers do not randomly assign themselves to housing agents or employers but select them for various reasons. Finally, different employees in the same establishment may behave differently. If a rental office has more than one agent who shows apartments, different experiences of the members of the pair may be traceable to differences in the behavior of the agent with whom they dealt.

Addressing the Limitations of Audit Studies

Ross and Yinger (2002) offer several options for addressing the limitations of audit studies. Three of the approaches they identify to address the problem of accuracy are (1) broaden the sampling frame to encompass methods other than newspaper advertisements (e.g., searching neighborhoods for rental or help-wanted signs); (2) examine whether the characteristics of the specific goods or services involved (e.g., housing unit) instead of the characteristics of the testers affect the probability of discrimination (Yinger, 1995); and (3) use actual characteristics—as opposed to assigned characteristics—of testers and determine whether controlling for these characteristics influences estimates of discrimination.

To address validity concerns, Ross and Yinger (2002) suggest a strategy of sending multiple pairs to each establishment, which would allow researchers to obtain the data needed to reduce the effects of the idiosyncratic characteristics of single pairs of testers. Testers could then be debriefed after each experience to determine the agent with whom they had dealt. Doing so would not remove the potential effect of different agents on the results obtained, but it would allow researchers to assess that effect. Use of additional pairs of testers would also address issues regarding the calculation of outcome measures. Using multiple pairs might help in distinguishing systematic from random behaviors of an establishment and should, at the very least, tighten the bounds one might calculate on the basis of different mathematical formulas. Of course, care would need to be taken to avoid sending so many pairs of confederates that the research would become obvious.

Another approach to addressing the limitations of omitted variables is to collect extensive information on the actual characteristics of testers, as opposed to assigning their characteristics, and to determine whether controlling for these characteristics influences estimates of discrimination. HUD's national audit study of housing discrimination, conducted in 2000, explicitly collected information on many actual characteristics of testers, such as their income (as opposed to the income assigned to them for the study), their education, and their experience in conducting tests.[6]

SUMMARY AND RECOMMENDATIONS

True experiments involve manipulation of the variable hypothesized to be causal, random assignment of participants to the experimental condition, and control of confounding variables. Experimental methods potentially provide the best solution to addressing causal inference (e.g., assigning disparate racial outcomes to discrimination per se) because well-designed and well-executed experiments have high levels of internal validity. In the language of contemporary statistics, experiments come closest to addressing the counterfactual question of how a person would have been treated but for his or her race, although they do not do so in a form that is easily translatable into direct measurement of the discriminatory effect.

[6]Results based on analyses of this information are available at http://www.huduser.org/publications/hsgfin/phase1.html [accessed August 19, 2003].

The experimental method faces challenges when applied to race, which cannot be randomly assigned to an actual person. Experimental researchers frequently manipulate racial cues (e.g., racial designations or photographs on a résumé) or train black and white confederates to respond in standard ways. In both approaches, an attempt is made to manipulate apparent race, while holding all other variables constant, and to elicit a response from the participants. Although the experimental method has uncovered many subtle yet powerful psychological mechanisms, a laboratory experiment does not address the generalizability or external validity of its effects. Therefore, it is unable to estimate what proportion of observed disparities is actually a function of discrimination.

Over the past two decades, laboratory experiments have focused more on measuring subtle forms of bias and nonverbal forms of discriminatory behavior and less on examining overt behaviors, such as assisting others. If laboratory studies were to be more focused on real-world-type behaviors, they could help analysts who use statistical models for developing causal inferences from observational data (see Chapter 7). Thus, the results of real-world-oriented laboratory studies could provide more fully fleshed-out theories of discriminatory mechanisms to guide the modeling work. In turn, real-world studies based on laboratory-developed theories could be usefully conducted to try to replicate, and thereby validate, laboratory results.

Because laboratory experiments have limited external validity, researchers turn to field experiments, which emphasize real-world generalizability but inevitably sacrifice some methodological precision. Field audit studies randomly assign experimental and control treatments (e.g., black and white apartment hunters) to units (e.g., a rental agency) and measure outcomes (e.g., number of apartments shown). Aggregated over many encounters and units of analysis, audit studies come closer than laboratory experiments to assessing levels of discrimination in a particular market. Both the accuracy and the validity of audit studies on discrimination have been questioned, however. Advocates of paired-testing and survey experiments have responded that all these limitations can be remedied.

Although generally limited to particular aspects of housing and labor markets (e.g., showing of apartments or houses and callbacks to job applicants), audit studies to measure racial discrimination in housing and employment have demonstrated useful results. It is likely that audit studies of racial discrimination in other domains (e.g., schooling and health care) could produce useful results as well, even though their use will undoubtedly present methodological challenges specific to each domain.

Recommendation 6.1. To enhance the contribution of laboratory experiments to measuring racial discrimination, public and private funding agencies and researchers should give priority to the following:

- Laboratory experiments that examine not only racially discriminatory attitudes but also discriminatory behavior. The results of such experiments could provide the theoretical basis for more accurate and complete statistical models of racial discrimination fit to observational data.

- Studies designed to test whether the results of laboratory experiments can be replicated in real-word settings with real-world data. Such studies can help establish the general applicability of laboratory findings.

Recommendation 6.2. Nationwide field audit studies of racially based housing discrimination, such as those implemented by the U.S. Department of Housing and Urban Development in 1977, 1989, and 2000, provide valuable data and should be continued.

Recommendation 6.3. Because properly designed and executed field audit studies can provide an important and useful means of measuring discrimination in various domains, public and private funding agencies should explore appropriately designed experiments for this purpose.

7

Statistical Analysis of Observational Data

Thus far we have made the case that randomized controlled experiments are the best approach available to researchers for drawing causal inferences. In the absence of experimental design, causal inference is more difficult. However, applying statistical models to observational data can be useful for understanding causal processes as well as for identifying basic facts about racial differences. Indeed, observational studies are the primary tool through which researchers have explored racial disparities and discrimination. The main goals of this chapter are to delineate the strengths and problems associated with measuring discrimination using observational studies and to identify methodological tools that are particularly promising for application in certain areas of research on discrimination.

We begin by discussing statistical decompositions of racial differences in outcomes using multivariate regressions. These decompositions are basically descriptive but are nevertheless an important tool for understanding what factors are related to observed differences as well as for measuring the magnitude of racial differences. In the next section, we continue with an outline of the fundamental issues that must be addressed to draw causal inferences about racial discrimination from statistical analyses of observational data. We illustrate the main issues by laying out a statistical model that can be used to measure discrimination in hiring decisions. As we see it, the hiring example is robust in the sense of surfacing all of the conceptual issues that hamper research on discrimination across domains, including the five domains on which we focus in this report (labor markets, education, housing, criminal justice, and health care). We discuss the strengths and limitations of existing approaches to measuring discrimination across these domains and suggest how approaches prevalent in one domain might usefully be employed in others.

Even a cursory review of the literature on labor markets, education, housing, criminal justice, and health care reveals that it is quite common for researchers to employ statistical models when addressing questions of racial discrimination (see Table 4-1). Given the range of domains we examine, we do not attempt to be exhaustive in our presentation. Instead, we provide examples from individual studies in particular domains to illustrate particular methodological issues. Our intent is to summarize what we see as the most important challenges that arise in using statistical models to study racial differences in outcomes. And although we make frequent use of labor market concepts as concrete examples throughout this chapter, the fundamental statistical issues underlie the measurement of discrimination in all domains.

It should be noted that the style of exposition in this chapter is more mathematical than that in the rest of the report. This mathematical presentation is necessary to make clear what statistical decompositions of racial differences measure. It is also needed for precision regarding the role of models as descriptions of the ways in which outcomes are determined in the presence of discrimination, the role of models and assumptions in drawing causal inferences regarding discrimination from observational data, the nature of the biases that arise when those assumptions are violated, and the ways in which alternative study designs can reduce those biases.

STATISTICAL ANALYSIS FOR RESEARCH AND LITIGATION

Before we proceed, a caveat is in order. This chapter attempts to illuminate state-of-the-art statistical methods that should be used by academic researchers attempting to detect the existence and magnitude of racial discrimination in a wide variety of domains. Statistical proof of racial discrimination may often be sought in other contexts in which the same degree of attention to methodological detail may be valued differently. In particular, courts are often called upon to decide discrimination cases in circumstances that are far less congenial to the detailed and sustained analysis of the academic researcher. Litigants often press for expert testimony based on something far short of state-of-the-art statistical practices that academic researchers might employ. In some instances, a straightforward analysis of the available data may appear to make a compelling case, but many outside the courts would argue for more details and alternative analyses to buttress the arguments.

In a paper commissioned for this panel, Nelson and Bennett (2003) investigate the courts' use of statistics to make decisions in cases alleging racial discrimination in employment. They analyze published federal court opinions on racial discrimination in employment that refer to "statistics" in some form. They compare practices in cases published in 2000–2002 (178 opinions) and in cases published in 1980–1982 (124 opinions) to evaluate changes over time. For cases published in 2000–2002, a preliminary analysis revealed that courts treat statistical data on racial discrimination conservatively; in other words, "they are reluctant to reject the null hypotheses of nondiscrimination and they are reluctant to hold that plaintiffs have met their burden of proof" (Nelson and Bennett, 2003:2). Similar results were found for the 1980–1982 period.

Most courts expect to see statistical evidence presented by plaintiffs in employment discrimination cases; Nelson and Bennett conclude, however, that in most cases courts are skeptical of statistics used to prove discrimination. Moreover, they do not appear very often to base opinions on statistical evidence in contrast to Supreme Court precedent or other judicial rulings. In fact, courts are relying on statistics less frequently now than they did in the 1980s, even though statistical techniques have improved. Moreover, the interpretation of statistical evidence varies across courts and cases.

Overall, the lack of credence given by courts to statistical evidence and the complexities of drawing inferences about racial discrimination from such data appear to be detrimental to plaintiffs. In both periods examined by Nelson and Bennett, plaintiffs lost to defendants more than three times to one, and it is becoming increasingly more difficult for plaintiffs to convince courts that their claims are valid.

The main reasons cited for not relying on statistical data in judicial opinions are (1) relatively small sample sizes, (2) difficulty in defining the comparison groups, (3) lack of relevant controls for nondiscriminatory explanations for disparities, and (4) the use of aggregated data across multiple job levels in a class action suit. These statistical issues (particularly the first three) were prominent in the cases examined within each time period.

Moreover, although there have been many sophisticated advances in statistical analyses, the analyses used in court cases typically involve simple comparisons between the racial composition of an applicant pool or a potential promotion pool and a set of selection outcomes (such as hiring or promotion). Few cases involve rigorous assessment of the use of multiple regression and other multivariate analyses. Courts discount small samples without considering the probabilities of outcomes, displaying a lack of statistical knowledge and reasoning. Courts also have no consistent approach to dealing with these problems. A recent Supreme Court decision, however (*Desert Palace v. Costa* [No. 02-679]), appears to open the door to expanded use of statistical methods to support inferences of discrimination in legal proceedings.

STATISTICAL DECOMPOSITIONS OF RACIAL DIFFERENCES

Two types of regression models have been used to decompose racial differences in outcomes. They are (1) regression models with race-specific intercepts, which assume that the effects of other variables (e.g., education) are similar for race groups, and (2) race-specific regression models that allow for interaction effects between race and other variables. All such models pose problems for interpretation.

Regression Models with Race-Specific Intercepts

A standard way to explore the difference in an outcome between groups is to decompose the difference into "explained" and "unexplained" components. To illustrate, suppose the researcher is comparing outcomes of white and black men. The simplest formulation is captured by the regression model

$$Y_i = \beta_0 + \gamma R_i + X_i \beta + u_i, \qquad (7.1)$$

where Y_i is the outcome of interest, such as a wage rate, with i indexing the individuals in the sample; R is an indicator variable that takes on the value 1 for blacks and 0 for whites; X_i is a set of variables that are believed to be relevant to the determination of Y_i; β_0 is the intercept; β is a vector of coefficients on the variables in X_i at a point in time; the coefficient γ captures the difference between groups in the average value of Y that is not accounted for by differences in X; and the error term u_i captures the effect of other factors that influence Y_i. The coefficients γ and β and u_i are defined so that the mean of u_i is unrelated to R_i and X_i. As we explain in detail below, unless the researcher is confident that he or she has measured all of the variables that are both correlated with R and relevant to the determination of Y, the model should be interpreted as descriptive rather than causal.

Let Y_b and X_b be the mean of Y and X for black men, and let Y_w and X_w be the mean of Y and X for white men. For concreteness, let Y be the wage rate. The average difference between the groups is

$$Y_w - Y_b. \qquad (7.2)$$

This difference can be decomposed as

$$Y_w - Y_b = \gamma + (X_w - X_b)\beta. \qquad (7.3)$$

The term $(X_w - X_b)\beta$ is the contribution of group differences in the observed characteristics X to the race gap in Y. For example, studies of the wage gap almost always include a measure of education among the variables in X. The product of the difference between whites and blacks in the education measure and the coefficient relating education to Y is the contribution of the education difference to the gap. The parameter γ is the portion of the group difference in the means of Y that is not accounted for by the difference in X_w and X_b given the weights β that the X variables have for a given time period. This parameter is a "catchall" that includes the effects of group differences in omitted factors that would influence Y in the absence of discrimination, as well as the effect of discrimination.

Everett and Wojtkiewicz (2002) provide a good example of this technique from the criminal justice literature and illustrate the fact that the technique is not restricted to linear regression models. They examine the racial and ethnic disparities in federal sentencing following implementation of guidelines that were intended to ameliorate past disparities in sentencing. They estimate ordinal logistic regressions to assess the relative odds of receiving a sentence within each successive quartile of the sentencing range for the committed offense. Adding various legal and extralegal factors to a baseline model including only indicators of race and ethnicity (black, Hispanic, Native American, and Asian, with white as the comparison group), they examine changes in the log-odds coefficient for each indicator. They find that significant racial differences in sentencing remain after accounting for legal factors (offense-related traits such as severity of the crime, as well as recidivism) and other, extralegal factors (such as age and education).

Race-Specific Regression Models

The above description covers a great deal of early research on discrimination that served as the basis for further work on measuring discrimination and explaining racial differences in a variety of social, political, and economic outcomes.[1] For the past 30 years, however, researchers have used a more general statistical model of such differences (or gaps) that allows for the possibility that the slope coefficients β differ between groups (e.g., an interaction between race and education; see Blinder, 1973; Duncan, 1968; Oaxaca, 1973). Suppose that Y_{wi} and Y_{bi} are determined by the equations

$$Y_i = \beta_{0w} + X_i \beta_w + u_i \text{ (for whites)}, \qquad (7.4)$$

$$Y_i = \beta_{0b} + X_i \beta_b + u_i \text{ (for blacks)}, \qquad (7.5)$$

where β_w and β_b are defined so that $E(u_{wi} \mid X_{wi}) = 0$ and $E(u_{bi} \mid X_{bi}) = 0$. Consequently, the means of Y_i for whites and blacks are

[1]The exposition in this section is based on Altonji and Blank (1999).

$$Y_w = \beta_{0w} + X_w \beta_w$$

and

$$Y_b = \beta_{0b} + X_b \beta_b,$$

respectively.

The difference in the mean of the outcome can be written as

$$Y_w - Y_b = (X_w - X_b) \beta_w + [(\beta_{0w} - \beta_{0b}) + X_b (\beta_w - \beta_b)]. \tag{7.6}$$

The first term in this decomposition is the portion of the total gap that is explained by average differences in characteristics of whites and blacks using the coefficients for whites (β_w) as the weights. In other words, it is the portion of the gap in Y that would be eliminated if the gap in X were closed and if the dependence of Y on X were the same for blacks and whites. The second term is the "unexplained" part of the gap in Y; that is, the difference that arises because the relationship between characteristics and outcomes, as summarized by the regression parameters including the difference in $(\beta_{0w} - \beta_{0b})$ in intercept terms, differs between groups. Blau and Beller (1992) offer a good example of such a decomposition (see Box 7-1).

Alternatively, the average outcome difference can be decomposed as

$$Y_w - Y_b = (X_w - X_b)\beta_b + [(\beta_{0w} - \beta_{0b}) + (\beta_w - \beta_b)X_w]. \tag{7.6'}$$

This alternative decomposition uses the coefficients from the model for blacks to determine the consequences for $Y_w - Y_b$ of the group differences in X and uses the mean of X for whites to determine the consequences of the difference $(\beta_w - \beta_b)$ in the slopes. The first term is the portion of the total gap that would be eliminated if the gap in X were closed and if the dependence of Y on X were the same for whites and blacks. The second term is the "unexplained" portion of the gap in Y. This second decomposition sometimes produces quite different results from those produced by the first. Many authors report both results or (occasionally) the average of the two (see Oaxaca and Ransom, 1999, for references).

Interpreting the Decomposition

The share of the total difference due to the second component in equation (7.6) is sometimes referred to as the "share due to discrimination." This is misleading terminology, however, because if any important control variables are omitted, one or more of the β coefficients, including the intercept, will be affected. The second component therefore captures both the

BOX 7-1
Use of Regression Models to Decompose Racial Differences

A study by Blau and Beller (1992) is a good example of the use of regression models to decompose differences between groups. They estimate forms of equations (7.4) and (7.5) for black and white men and women in various experience categories, where experience refers to number of years since leaving school. They estimate two sets of regressions. The first uses the logarithm of earnings as the dependent variable and includes measures of education, potential labor market experience and its square, the natural log of annual weeks worked, dummy variables for part-time work status, veteran status (in the case of males), marital status, and dummy variables for three regions and urban residence. The regressions for women include controls for the number of children. The authors also report decompositions based on a second set of regressions that adds a list of dummy variables for major occupation category and for employment in the government sector to the regressions.

Blau and Beller (1992:Table 3) report results of their decompositions using the coefficients from the regression model estimated on the white sample. They present separate results for 1971, 1981, and 1988 and use the results for these three periods to investigate changes between 1971 and 1981 and 1981 and 1988. When occupation dummies are excluded, the 1971 results for males with 10 to 19 years of potential experience estimate that 0.209 of the total gap of 0.452 in the log of earnings is due to differences in the means of the observed characteristics that determine earnings. Blau and Beller report that, of the part of the gap that is due to differences in observed characteristics, 0.096 is due to differences in the means of education and 0.061 is due to differences in the means of the variables that measure work hours. They find that 0.243 of the total gap of 0.452 is "unexplained" and reflects differences in the intercepts and slope coefficients on the variables in the earnings model. The results for 1988 show an increase in the total gap to 0.505. The explained gap rises to 0.331, primarily because of an increase in the portion of the gap associated with hours worked during the year, while the unexplained gap actually falls.

effects of discrimination and the unobserved group differences in factors that would be expected to determine Y in the absence of discrimination. If there is a gap in favor of whites (blacks) in most of the omitted variables that boost Y, the second component will tend to overestimate (underestimate) the effects of discrimination. On the other hand, omitted variables that are correlated with X will influence the coefficients β, potentially causing the "unexplained" portion of the gap to be either an overestimate or an underestimate of discrimination. Finally, the inclusion in X of variables that are themselves an outcome in a particular domain, such as occupation or position within a firm in a study of earnings differences, may cause the second component to underestimate or less often overestimate discrimination.

It is also misleading to label only the second component as the result of discrimination. This is the case because discriminatory barriers in the labor market and elsewhere in the economy can affect the X variables, which are the characteristics of individuals that matter in the labor market. We discuss this point more below and in Chapter 11.

Although one must be mindful of the limits of what can be learned from equation (7.6), it is nevertheless a simple and powerful way to summarize information on some of the factors that underlie group differences. The decomposition analysis can be extended to study change in group differences over time (see Box 7-2).

Two Pitfalls in Statistical Decomposition

Many researchers further decompose the "explained" gap into the contribution of subgroups of variables. For example, suppose that X contains sets of indicators for region of the country, for city size, and for educational attainment (less than high school, some high school, high school, some college, college, and some graduate school). One would like to know the contribution of each of these sets of indicators to the explained and unexplained portions of the gap. The contributions of each set of variables to the explained gap are identified and can be estimated separately. The problem, however, is that the contributions of the individual variables to the unexplained gap are not identified separately and depend on the choice of the reference category for each variable. That is, one cannot distinguish the contribution to the overall unexplained gap of racial differences in the coefficients on region of the country from the contribution of racial differences in the coefficients on city size. See Jones (1983) and especially Oaxaca and Ransom (1999) for a more extended discussion of this issue and citations of a number of studies that have included such detailed decompositions.

When the relationship between Y and the X variables is highly nonlinear and the racial difference in the distribution of X is large, a lack of overlap between the black and white distributions in the X variables may make it difficult or impossible to estimate the decompositions of equations (7.6) or (7.6′) reliably. This problem may not be obvious if researchers use functional form specifications of equations (7.4) and (7.5) that are not sensitive to potential nonlinearities. Barsky et al. (2002) and Altonji and Doraszelski (2002) investigate the problem posed by lack of overlap in black

BOX 7-2
Statistical Decompositions over Time

One way to analyze the sources of change over time in the outcomes of various groups is to differentiate between periods. The simplest way to proceed is to perform decompositions in two different years and compute the change in the "explained" component $[(X_{wt} - X_{bt}) \beta_{wt}]$ and the change in the "unexplained" component $[(\beta_{0wt} - \beta_{0bt}) + (\beta_{wt} - \beta_{bt})X_{bt}]$, where we introduce t as a time subscript to make explicit the fact that the equations refer to a particular year. Blau and Beller (1992) and many other studies do this. However, the change in each of the two components combines the effects of changes in the race gap in characteristics and in the race gap in coefficients.

A more detailed decomposition of change over time can be obtained as follows. Let the operator Δ represent the average difference between members of group 1 and group 2 in a particular year. For concreteness, let the outcome Y denote the wage rate. The change in wage differentials between time periods t' and t can be expressed as

$$\Delta Y_{t'} - \Delta Y_t = (\Delta X_{t'} - \Delta X_t) \beta_{wt'} + \Delta X_t(\beta_{wt'} - \beta_{wt}) + [(\Delta\beta_{0t'} - \Delta\beta_{0t}) + (\Delta\beta_{t'} - \Delta\beta_t) X_{bt'}] + (X_{bt'} - X_{bt})\Delta\beta_t.$$

The first term on the right-hand side of the equation represents the contribution of the relative changes over time in the observed characteristics of the two groups to the change between t and t' in the wage gap. The second term is the effect of changes over time in the coefficients for group 1, holding differences in observed characteristics fixed. These first and second terms' two factors capture the change over time in the explained portion of the wage gap that would be expected given changes in the characteristics of the two groups and the coefficients on those characteristics for whites in periods t and t'.

The third and fourth terms capture the change in the unexplained component of the gap, $(\beta_{wt} - \beta_{bt}) X_{bt}$ in equation (7.4). The third term is the effect of changes over time in the gap in the coefficients between the two groups. The fourth term accounts for the fact that changes over time in the characteristics of group 2 alter the consequences of differences in group coefficients $(\beta_{wt} - \beta_{bt})$. Researchers typically compute each of these terms, as well as the subcomponents corresponding to individual elements of X and β.

A limitation of this decomposition is that it does not provide much insight into how the wage gap is affected by changes in the overall wage distribution, such as occurred over the 1980s when the returns to skill rose rapidly. Increases in the dispersion of wages will increase the gap between the mean wages of whites and blacks (given that whites are above the mean and blacks below), even if there is no change in the skill distributions of whites relative to blacks or in the level of discrimination. Juhn et al. (1991) and Card and Lemieux (1994, 1996) suggest ways to isolate the effect of a change in the dispersion of the unobservable wage components affecting both groups from a change in the location of the skill distribution of group 2 relative to group 1.

Altonji and Blank (1999) provide a detailed discussion of the methods used in these papers. A brief summary of Juhn et al.'s (1991) basic results indicates what one can learn from their type of analysis. Using data from the Current Population Survey, they find that, between 1979 and 1987, changes in levels of education and experience reduced the black–white wage gap (in logs) for men by 0.34 (black characteristics moved closer to white characteristics), whereas increases in the returns to education and experience increased the gap by 0.27. They find that 0.33 of the 0.34 unexplained widening in the wage gap can be attributed to changing wage inequality affecting both whites and blacks. In short, they find that relative wages of blacks declined because black men were disproportionately located at the lower end of an increasingly unequal wage distribution.

An alternative approach, used by Murnane et al. (1995), is to examine the sensitivity of estimates for 1978 and 1986 of the unexplained race gap in earnings to adding more detailed cognitive measures, such as test scores, to earnings regressions. They find a smaller race gap in the 1970s that is less sensitive to inclusion of test scores, particularly for males. This result is broadly consistent with the analysis in Juhn et al. (1991).*

---

*For an alternative view of this evidence, see Darity and Mason (1998).

and white income distributions in the context of studies of the black–white wealth gap, with differing conclusions. Barsky et al. (2002) standardize for the effects of income by reweighting the sample of whites to have the same income distribution as the sample of blacks.[2]

Summary: Decomposition and Residual "Effects" as Racial Discrimination

The use of multivariate regression and related techniques to decompose racial differences in some outcome of interest into a portion due to differences in the distribution of observed characteristics and a portion not explained by those characteristics is an essential tool for describing racial differences. The most informative studies use explanatory variables that both measure the most important determinants of the outcome under study and are likely to have different distributions by race. But the residual race differential may include not only any effect of discrimination but also the effect of other omitted factors that would generate different outcomes by race even in the absence of discrimination. Hence the unexplained gap may overestimate or underestimate the effects of discrimination.

INFERRING DISCRIMINATION FROM STATISTICAL ANALYSIS OF OBSERVATIONAL DATA

In this section, we discuss some of the more frequently encountered obstacles to causal inference in statistical studies of racial discrimination. As discussed in Chapters 5 and 6, it could be relatively easy to estimate the degree of discrimination if only it were possible to manipulate a person's race. Except in very limited or special circumstances (e.g., an audit framework), race cannot be randomly assigned.[3] Statistical models are widely used in observational studies in an attempt to replace the experimental control that could ensure an "all-else-equal" comparison. Again, the crucial problem that must be addressed to draw a causal inference from observa-

[2]See also DiNardo et al. (1996). One can allow for arbitrary nonlinearity in the relationship between the outcome and X by first estimating a statistical model of the probability that a person is black as a function of X and then matching whites and blacks on the basis of this probability, which is called the "propensity score." Nopo (2002) uses a nonparametric matching technique to decompose the gender gap in wages in Peru. Black et al. (2002) use a nonparametric matching technique to estimate the fraction of the earnings differences among college-educated white, black, Hispanic, and gay men that is explained by age, specific college major and degrees, English-language proficiency, family background, and region of birth.

[3]Even in an audit framework, random assignment is not simple. One of the key controversies in audit studies is the extent to which the designs can approach the classic random assignment paradigm (Heckman, 1998). See the discussion in Chapter 6.

tional data is that the researcher has no control over which subjects have which attributes. Essentially, the inference that race has a causal effect on an outcome (because of racial discrimination) is drawn by shaping a set of statistical correlations using other information and assumptions formalized in a model of how the process under study is determined. This approach is typical of statistical analysis of observational data and is not unique to the problem of discrimination. Sometimes, the most we can claim is that the evidence is consistent with a certain explanation, with the caveat that other plausible explanations cannot be excluded.

Below we identify and discuss common obstacles to causal inference and some of the solutions proposed in the literature. We begin with a brief introduction to the essential role of theoretically informed models and adequate data in drawing causal inferences from observational data. We then illustrate this point with an extended example involving hiring decisions in the labor market. Finally, we discuss two of the most important sources of bias in observation studies of discrimination—omitted variables bias and sample selection bias.

Developing Statistical Models

According to Sir Ronald Fisher, as quoted by Cochran (1965:252), "When asked in a meeting what can be done in observational studies to clarify the step from association to causation, [I] replied: 'Make your theories elaborate.'"[4] To justify causal inference from observational data, we need a theoretically informed model that depicts, as accurately as possible, the specific process in which we are attempting to assess the presence and magnitude of racial discrimination. Depending on the particular process and context, one may have more or less information on which to base a theoretical model and then translate it into a statistical model. Laboratory experiments (see Chapter 6) are designed precisely to test the plausibility of various detailed theoretical frameworks.

As discussed in detail in Chapter 5, there is a growing literature that formalizes the assumptions and the deductive process needed to draw cause-and-effect inferences from statistical data. The key idea underlying this literature is the hypothetical counterfactual introduced in Chapter 5: What would have happened if the applicant for a job or rental housing had been white rather than nonwhite but nothing else had changed? Obviously, the counterfactual situation cannot be observed and compared with what actually occurred. Therefore, to draw a causal inference from experimental or observational data, it is necessary to specify assumptions and conditions

[4]This discussion draws on Rosenbaum (2001).

under which the counterfactual logic can be applied. Assumptions from the causal literature are particularly important for justifying the use of regression methods for drawing causal inferences. To draw inferences from running regressions on observational data, substantial prior knowledge about the mechanisms that generated the data must be used to support the necessary assumptions. Studies vary substantially in the degree to which the necessary assumptions are adequately justified. Below we discuss some of the specific issues that must be addressed in such models and their assumptions to draw causal inferences.

Example: Hiring Decisions in the Labor Market

In this section, we lay out a generic framework that underlies many statistical approaches to measuring discrimination. The example is from the labor market domain—in particular, hiring. However, the principles described here are quite general, and the issues raised apply across various domains. Our main purpose is to identify what a researcher must know about how an outcome is determined and what data must be available if discrimination is to be measured. We also discuss some of the most important limitations and issues of interpretation surrounding statistical studies of race differences based on observational data.

To set the stage, suppose the researcher is interested in understanding an outcome variable, labeled y. In the labor market context, y might be the probability of getting hired by a particular firm for a particular job. In other contexts, such as housing or education, y might be the probability that a housing loan application will be approved or that a person will be admitted to a university. To develop an adequate model of the phenomenon, the analyst needs to have a good understanding of the process that would determine y in both the absence and the presence of discrimination. In this example, the researcher would want information about the legitimate criteria (e.g., education or experience) used by the firm's recruiters to screen applicants for the position.

In the case of hiring, a rational, profit-maximizing, nondiscriminating firm would prefer to hire people who are the best suited to perform well in the jobs for which they are screening applicants. Let the variable P denote productivity in a particular position. To make the model of the decision process as realistic as possible, we distinguish among the variables that determine productivity on the basis of whether they are known to the researcher, the employer, both, or neither. Let X_1 be the factors that are known to both researcher and employer. Examples of X_1 factors might be years of education or labor force experience, or other criteria that are easily observable from an application form. Let X_2 be a set of factors known to the employer but not to the researcher. Examples of X_2 might be such factors as the performance of the applicant in an interview, which is likely to affect screeners' hiring decisions but unlikely to be observed by the researcher. Let Z be a set of variables known to the researcher but not to the employer. Such variables might comprise information collected as part of a survey of the applicants by the researcher. For example, the researcher but not the employer might know that the applicant's spouse works at the company. Finally, even the most diligent employers and researchers might not have access to all the factors likely to affect a person's productivity in a given job. Let Q be a set of factors that affect productivity but are observed by neither the employer nor the researcher.

For simplicity, we assume throughout that all relationships among the variables are linear,[5] so that X_1, X_2, Z, and Q determine productivity according to

$$P = X_1B_1 + X_2B_2 + ZG_1 + QG_2, \qquad (7.7)$$

where B_1, B_2, G_1, and G_2 are weights capturing the importance of each set of factors in determining productivity in the job. For example, if the factors that are unknown to both the employer and the researcher (i.e., Q) are not very important, G_2 will approach zero, and QG_2 will have little affect on productivity P. Similarly, if the Z factors are not very important as determinants of productivity, the weight G_1 will approach zero.

Two important points must be made about the framework summarized in equation (7.7). First, because this equation is a model of hiring in the absence of discrimination, we exclude the race (labeled R) of the individual from the list of X_1 factors, despite the fact that R might easily be observed by both the firm and the researcher. In this model, therefore, R has no effect on productivity, which is fully determined by X_1, X_2, Z, and Q. We relax this assumption below in discussing a case in which race does influence productivity as viewed by the firm because its customers are prejudiced.

Second, equation (7.7) has the virtue of specifying precisely what the decision criteria would be for a rational firm seeking to hire the most productive candidates. In particular, a rational firm will base its hiring decisions on the expected productivity of the applicants, given the information it has, X_1 and X_2. If the firm uses only X_1 and X_2 to make its productivity assessment, ignoring race, the firm's hiring decision will be a function of the expected value of P given X_1 and X_2. The firm's expectation of productivity conditional on the information it has (X_1 and X_2) can be denoted

[5]The linearity assumption in equation (7.7) is a strong one, which we make to simplify the situation sufficiently for pedagogical purposes. For example, with binary variables as outcomes, equation (7.7) will typically not apply in this form and the analysis will require modification.

$E(P \mid X_1,X_2)$, which is a function of the information the firm has (X_1 and X_2), taking into account the expected value of the information it does not have (Z and Q). The expected values of Z and Q, conditional on X_1 and X_2, are $E(Z \mid X_1,X_2)$ and $E(Q \mid X_1,X_2)$, respectively. After weighting each of these terms by its importance as a determinant of productivity (B_1, B_2, G_1, and G_2, respectively), the rational firm's expectation of productivity will be

$$E(P \mid X_1,X_2) = X_1B_1 + X_2B_2 + E(Z \mid X_1,X_2)G_1 + E(Q \mid X_1,X_2)G_2. \tag{7.8}$$

We assume throughout that all conditional expectations are linear functions, in which case equation (7.8) can be rewritten as

$$E(P \mid X_1,X_2) = X_1B_1^* + X_2B_2^*, \tag{7.9}$$

where

$$B_1^* = B_1 + \pi_1,\ B_2^* = B_2 + \pi_2,$$

and the equation

$$E(Z \mid X_1,X_2)G_1 + E(Q \mid X_1,X_2)G_2 = X_1\pi_1 + X_2\pi_2$$

defines π_1 and π_2.

The intuition behind π_1 and π_2 can be seen by rewriting equation (7.9) as

$$E(P \mid X_1,X_2) = X_1(B_1 + \pi_1) + X_2(B_2 + \pi_2).$$

The π_1 and π_2 terms, respectively, capture the ways in which X_1 and X_2 affect the firm's estimate of productivity indirectly via their associations with the unobserved factors Q and Z.

We now turn to the hiring decision itself. For a rational, nondiscriminating firm, the hire probability is an increasing function of $E(P \mid X_1,X_2)$. For simplicity, we assume that the relationship is linear, in which case

$$y = \text{constant} + a_1[E(P \mid X_1,X_2)] + u;\ a_1 > 0, \tag{7.10}$$

which, using equation (7.9) to substitute for $[E(P \mid X_1,X_2)]$, can be written as

$$y = \text{constant} + X_1(a_1B_1^*) + X_2(a_1B_2^*) + u. \tag{7.11}$$

The error term u captures random noise in hiring, as well as the fact that whether an individual with a particular set of characteristics will be hired at a given point in time will be affected by random variation in the quality of the other applicants at that time. We assume that both of these sources of variation are unrelated to race (R).

For a nondiscriminating firm, note that there is no role for R in the hiring equation even if R happens to be correlated with Z or Q and thus correlated with P. The reason is that the hiring decision y depends on $E(P \mid X_1,X_2)$, not on $E(P \mid X_1,X_2,Z,Q)$ or P itself. (The firm does not observe Z, Q, or P.) If R is added to the model of equation (7.11), it will enter with a coefficient of zero, and the researcher will typically find that for a nondiscriminating firm there is no evidence of a difference in the hiring rates of members of different racial groups who have the same values of X_1 and X_2. (To focus on the key ideas, we assume throughout this section that samples are large enough that we can ignore sampling error in estimates.)

Figure 7-1 depicts the model of the nondiscriminating firm. The arrows from the box containing the firm's information (X_1 and X_2) and other factors (Z, Q) to productivity capture the fact that all four variables determine productivity. However, there is no arrow from the box containing (Z, Q) to the firm's judgment about productivity or to the hiring outcome because the firm does not observe Z or Q. For the same reason, there is no arrow linking actual productivity P to the firm's judgment about P or to the outcome. Race can be correlated with X_1, X_2, Z, and Q, but the nondiscriminating firm makes no use of race in making a judgment about productivity or in deciding the outcome. Consequently, there is no arrow from race to the outcome.

Now we allow for the possibility that the firm discriminates and bases its decisions on both expected productivity and race. A simple way to capture this possibility is with the hiring rule

$$y = \text{constant} + (1 - \alpha)a_1 E(P \mid X_1,X_2) + \alpha R + u, 0 \quad \alpha \quad 1, \qquad (7.12)$$

where α is the relative weight placed on race by the firm's screeners, and $1 - \alpha$ is the relative weight on productivity. This can be rewritten as

$$y = \text{constant} + X_1(\alpha' B_1^* a_1) + X_2(B_2^* a_1) + \alpha R + u, \qquad (7.13)$$

where $\alpha' = (1 - \alpha)$. The parameter α is the difference in the hiring probability that is due to discrimination on the part of the firm. In contrast to the situation for a nondiscriminating firm, when adding R to equation (7.11) will yield a coefficient of zero, α will not be zero for a discriminatory firm when estimating equation (7.13). Figure 7-2 shows the model of a discriminating firm. In contrast with Figure 7-1, Figure 7-2 introduces an arrow

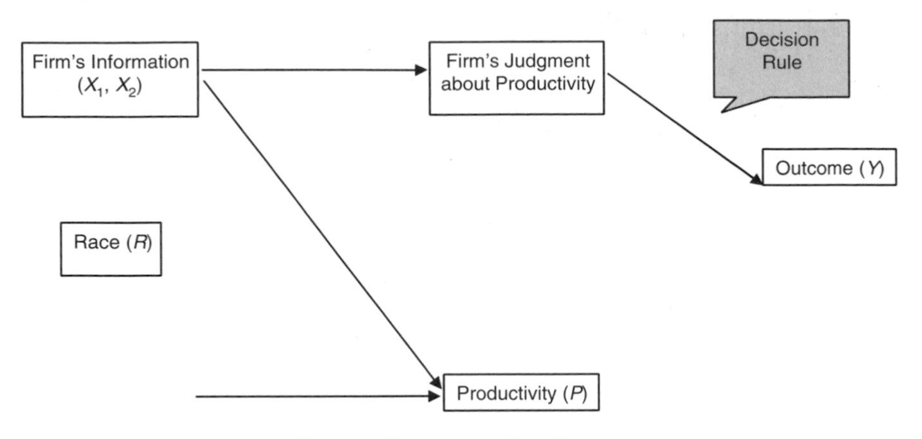

FIGURE 7-1 Model of a nondiscriminating firm.

from race to the outcome Y that represents the influence of discrimination. In this model, the strength of the link is α.

What should be obvious from this discussion is that to correctly estimate α, the researcher must have quite a bit of knowledge about how the firm behaves. First, the researcher must have a solid understanding of how the firm would behave in the absence of discrimination. The above model assumes that a nondiscriminating firm would hire on the basis of expected productivity in the firm. This presumption should guide the search for control variables (X_1, X_2). Second, the researcher must know how the firm

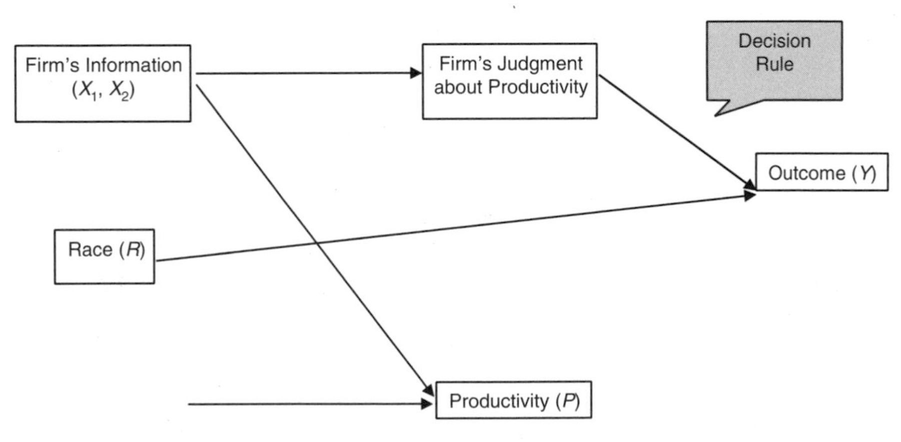

FIGURE 7-2 Model of a discriminating firm.

predicts productivity and must have data used by the firm. In the notation of the model, there must not be any X_2 variables—variables used by the firm of which the researcher is ignorant. Third, one must know that the firm does not use Z in its hiring decision. Otherwise, if Z were correlated with race and data on productivity were available, it would be easy to draw the wrong inference about α from a joint analysis of productivity and hiring decisions even if there were no omitted X_2 variables. (For example, the researcher might have data on performance on tests that were administered as part of a survey and would not be observed by firms.) A similar point applies to other variables the researcher happens to observe that have no relationship with P conditional on X_1, X_2, Z, and Q. Fourth, one must know enough about the relationship between P and X_1 and between y and $E(P \mid X_1)$ to be able to specify a functional form relating y to X_1 that is both a good approximation and estimable given the data at hand. In the above discussion, we have specified the relationships to be linear in the variables (which may include nonlinear transformations of a set of underlying variables). Even if they are in fact linear, the researcher will typically not know this to be the case and may have to use flexible functional forms or matching techniques or both.

What would the sources of such knowledge be? To continue with the hiring example, in some relatively rare situations the researcher may have deep knowledge of how hiring decisions are made and have access to nearly the same information as the firm (see the example in Box 7-3). In other cases, the researcher may have general knowledge of the most important determinants of P, based on, for example, case studies of similar jobs or interviews with employers or employees, or other knowledge about how the labor market works that is relevant to the problem. When data on P as well as y are available, the researcher can estimate the relationship between P and X_1 and so can draw inferences with weaker assumptions.[6]

Munnell et al.'s (1996) study of discrimination in mortgage lending is an analysis in which the quality of the data used and the level of understanding of how outcomes should be determined in the absence of discrimination are sufficiently high that the results are informative about discrimination in mortgage markets. Munnell et al. previously interviewed lenders to identify the factors important to them in determining the suitability of an applicant for a loan, and, thus, their choice of variables to include in equations (7.9) and (7.13) was well motivated. They make a good case that they

[6]The Multi-City Study of Urban Inequality, analyzed, for example, by Holzer (1996) and Coleman et al. (2002), contains information about skill requirements, worker characteristics, wages, and performance for the most recent hire for a sample of employers. For a subset of cases, the data can be matched to a household survey, which contains additional information on employees.

BOX 7-3
Gender Segregation of Jobs

A study of gender segregation of jobs by Fernandez and Sosa (2003) is one of very few analyses of how an organization selects individuals from a pool of applicants (other examples include Fernandez and Weinberg, 1997; Fernandez et al., 2000; and Peterson et al., 2000). The Fernandez and Sosa study is unusual for the level of detailed information gathered on the hiring and screening practices in the setting being examined. The authors interviewed company personnel about the specific criteria they used for screening applicants for a specific entry-level job—customer service representative. They then coded nearly 4,300 original paper applications received for that job over a 2-year period to reflect the concerns of the personnel screening applicants. Screeners said they looked for evidence of past experience in financial services or customer service settings. They also preferred people who were employed at the time of application, although they said they shunned "overpaid" people who they feared would be likely to leave prematurely. Screeners also said that they placed relatively little weight on formal education when screening individuals. However, they avoided "overeducated" people because they feared such people would leave quickly.

Fernandez and Sosa found that the application pool was two-thirds female, a ratio that roughly matched the gender composition of customer service representatives at the start of the study. Over time during the study, the percentage of women increased with each successive step in the screening process, to 69 percent of interviewees, 77 percent of offers, and 78 percent of hires. According to the criteria recruiters said they found desirable, female applicants were better qualified than males at the time of application. Using these criteria, Fernandez and Sosa then developed predictive models of who would be interviewed and who would be offered a job, conditional on being interviewed. Controlling for the applicant characteristics available from the application material did not fully explain screeners' apparent preference for hiring females.

Although the degree of knowledge available about the hiring process in this study is considerable, it is not complete. For example, the authors did not have access to data reflecting candidates' performance during interpersonal interactions with screeners, either on the telephone or in person. However, the interviews with the screening personnel and the detailed data collected from the application forms allow for a much closer match between the statistical models used and the process under study than is typical in observational studies.

have measured the most important factors considered by lenders. They show that differences between blacks and whites and Hispanics and whites in such characteristics as income explain much of the higher rejection rates for minorities, but they also find that a substantial unexplained gap remains. Although some have raised questions about the study (see Harrison, 1998), the authors provide a strong foundation for their conclusion that "a serious [racial discrimination] problem may exist in the market for mortgage loans" (Munnell et al., 1996:51).

PROBLEMS WITH MEASURING DISCRIMINATION BY FITTING STATISTICAL MODELS TO OBSERVATIONAL DATA

In addition to the concept of manipulability discussed in Chapter 5, any causal analysis that fits multiple regression models to observational data must address several issues. One such issue is whether the structure of the regression model—the variables and their functional form—captures the causal mechanism with sufficient accuracy. If, as is frequently the case, the variables or their functional form are not specified in advance of the analysis, there is a danger of overfitting. Also, a purely exploratory "fishing expedition" analysis may not be replicable. To address these issues, variables not appearing in a model but highly correlated with model variables could be substituted to evaluate whether alternative models based on these variables would fit as well. Also, cross-validation by examining the performance of the model on a subset of the data or on another data set could improve confidence in the robustness of the inference. A regression model would be suspect if the coefficients changed drastically when the model was fit to a subset of the data. How to address these generic issues is described in many texts on multiple regression (e.g., Weisberg, 1985).

There are two key limitations of statistical analysis of racial differences based on observational data that we discuss in more depth using the above framework. The first and most important of these is omitted variables bias; the second is sample selection bias.

Omitted Variables Bias

Before we discuss omitted variables bias in the context of the above model, we provide in Box 7-4 a simple illustration of this type of bias involving a study of graduate school admissions (Bickel and O'Connell, 1975), which is an example of what has come to be called "Simpson's paradox." The hypothetical data in the example involve a case in which the gender disparity in admissions reverses sign when an additional relevant variable is introduced—hence, the term omitted variable bias. Precisely the same phenomenon can occur in regression studies of racial disparities.

BOX 7-4
Simpson's Paradox

Consider a fictitious college, comprising four academic departments, referred to as A, B, C, and D. Members of two different racial groups apply to this college. The following admissions decisions are made, aggregated by the racial group of the applicant:

	Department A		Department B		Department C		Department D		Total	
Racial Group	Adm.	Rej.	Adm.	Rej.	Adm.	Rej.	Adm.	Rej.	Adm.	Rej.
I	35	25	30	30	5	0	15	5	85	60
II	5	5	5	10	50	10	40	25	100	50

NOTE: Adm., admitted; Rej., rejected.

Clearly, it is relatively more difficult to get accepted by departments A and B as compared with departments C and D. Looking at departments individually, we find that the admissions rate for individuals from the two racial groups are as follows:

Racial Group	Department A	Department B	Department C	Department D
I	0.58	0.50	1.00	0.75
II	0.50	0.33	0.83	0.62

Clearly within each department, individuals from racial group I have a higher admissions rate than those from racial group II. Aggregating the data across departments, however, the college-wide admission rates are 0.59 for individuals from racial group I and 0.67 for individuals from racial

Omitted variables bias poses a serious problem for the large share of studies of racial differences in surveys (e.g., the Current Population Survey or decennial census long-form sample) having only a limited set of the characteristics that may reasonably factor into the processes under study. In such circumstances, it is possible to measure the extent of the difference in the outcome that is associated with race, but it is not possible to decompose

group II, reversing the pattern uniformly exhibited at the department level. This apparent paradox is referred to as Simpson's paradox and is a simple illustration of the problems that can be faced when examining aggregate data on discrimination. The explanation for this seeming paradox is that the aggregate analysis implicitly assumes that the racial groups are relatively homogeneous with respect to the propensity to apply to the various departments at the college and that the departments are relatively homogeneous with respect to admission rates, but these assumptions do not obtain. A good experimental design—which would not be possible to carry out—might have avoided this problem by restricting the racial groups to applying equally to the four departments. This balancing across departments has not occurred, however. Instead, the aggregate analysis for racial group I is weighted by department as follows: 0.41, 0.41, 0.03, and 0.14. In comparison, the aggregate analysis for racial group II is weighted as follows: 0.07, 0.10, 0.40, and 0.43. This imbalance, in conjunction with the varying admission rates of the departments, has caused the aggregate analysis to be misleading. In that analysis, department is functioning as a hidden variable that is correlated both with admission and with racial group.

More informally, what has happened is that racial group I has applied much more often to departments that have lower admission rates, and the reverse has occurred for racial group II. So the aggregate analysis is, roughly speaking, comparing the admission rates for racial group I in departments A and B with the admission rates for racial group II in departments C and D, which is an unfair comparison.

An obvious follow-up question is how one knows whether the same mistake is being made in the department-level analyses with respect to improper aggregation of disaggregated analyses. For instance, what if we make the four 2 × 2 tables into eight 2 × 2 tables by splitting according to whether applicants are in state or not in state? Could the findings switch back again? The answer is that we cannot say without further analysis. This is why the search for relevant hidden variables is so crucial. More generally, this example demonstrates the complexity and subtlety of analyses of the presence of discrimination and the need to carefully scrutinize statistical models used for this purpose.

that difference into a portion that reflects discrimination and a portion that reflects the association between race and omitted variables that also affect the outcome. Because researchers generally do not know whether or what critical variables have been omitted, they must be very careful in making the leap from statistical decompositions in a statistical accounting exercise as discussed above to conclusions about the role of discrimination.

Turning back to the specific regression model of hiring outlined above, if elements of X_2 are correlated with R and are important for productivity, failure to control for them when estimating the discrimination parameter α will lead to bias in estimates of the extent to which race has an independent effect on hiring probabilities. The model laid out above in equations 7.7 through 7.13, however, allows us to be more precise about ways in which the omitted variables threaten an inference about discrimination. The discussion below provides a foundation for possible solutions to the problem of omitted variables bias.

Omitted variables affect the estimation of α as follows.[7] The researcher attempts to estimate α by regressing y on X_1 and R. Given equation (7.12), the coefficients of the regression are estimates of the regression model

$$y = \text{constant} + \alpha' a_1 E[E(P \mid X_1, X_2) \mid X_1, R] + \alpha R + u^*,$$

where u^* is uncorrelated with X_1 and R by definition of

$$E[E(P \mid X_1, X_2) \mid X_1, R]$$

and because of the properties of the error term u in equations (7.11), (7.12), and (7.13).

Because we are assuming that expectations are linear, we can write

$$E[E(P \mid X_1, X_2) \mid X_1, R] = X_1(B_1^* + \gamma_1) + R\gamma_2,$$

where γ_1 and γ_2 are the coefficients relating the mean of $X_2 B_2^*$ to X_1 and R, respectively. Then the regression of y on X_1 and R is

$$\begin{aligned} y = \text{constant} &+ X_1(B_1^* + \gamma_1)a_1\alpha' + \\ &R[(\gamma_2 a_1)\alpha' + \alpha] + u^*, \end{aligned} \tag{7.14}$$

where u^* is uncorrelated with R and X_1. The bias in the coefficient on R as an estimator of α is $\gamma_2 a_1 \alpha'$. The bias is due to correlation between R and the omitted variables X_2. The bias will be positive if R is positively related to the index $X_2 B_2^*$ of omitted factors that raise productivity, controlling for X_1; it will be negative if the opposite is true.

Sample Selection Bias

Sample selection bias exists when, instead of simply missing information on characteristics important to the process under study, the researcher

[7] For simplicity and to keep the emphasis on the most likely case, we ignore the fact that the researcher may have some additional information in Z that is unknown to the firm but possibly correlated with X_2. (In most circumstances, the researcher will not know much that is relevant for P that the firm does not also know.)

is also systematically (i.e., nonrandomly) missing subjects whose characteristics vary from those of the individuals represented in the data. The classic example pertains to the study of the wages people could expect to get in the labor market (see Heckman, 1979). In that context, the selection problem arises because wages are observed only for those who enter the labor market and find employment. People who have sought and managed to find work are likely to have better labor market opportunities than those who are observed not to be working.

In the context of our example, discrimination in hiring lowers the incentives of minorities with a given value of X_1 and X_2 to apply to a firm. Suppose that data are available only on applicants rather than on the entire pool of potential applicants. If the researcher observes all the information the firm uses in making its hiring decisions (X_2 has no elements), and if applicants know this is the information the firm uses, in addition to R, to make its decisions, then selection in who applies will not lead to bias when one uses equation (7.14) to estimate α. This outcome is extremely rare. Intuitively, discrimination may lower the probability that a black individual with a given value of X will apply, but estimation of equation (7.14) will still uncover the race difference in the hire probability between blacks and whites who are otherwise identical in the relevant characteristics.

Unfortunately, researchers seldom have as much information on applicant qualifications as the firm, in which case X_2 has elements, such as performance on an application test or in an interview. Then in the presence of discrimination, minorities with more favorable values of X_2 will choose to apply more frequently, conditional on X_1. Consequently, selection will tend to induce a positive relationship between R and the index $X_2 B_2^* a_1$ that influences hiring but is omitted from the regression. (Again, it is omitted because the researcher *does not* observe X_2.) Estimates of the amount of discrimination on the basis of regressions of y on X_1 and R will be understated as a result.

It is also important to note that, if discrimination in hiring influences the applicant pool, one cannot infer the consequences of discrimination in hiring from knowledge of α alone. One would need to know the effect of discrimination at the hiring stage on the racial mix of who applies.

POSSIBLE SOLUTIONS TO PROBLEMS OF USING STATISTICAL MODELS TO INFER DISCRIMINATION

There are many situations in the social sciences in which the researcher is confronted with an omitted variables problem that is parallel to that discussed above. The most common approach for dealing with omitted variables bias is to use an instrumental variables estimator, which amounts to isolating the relationship between the outcome and a particular source of variation in the explanatory variable of interest that is unrelated to the omitted factors. This strategy is not likely to be available in observational studies in the case of race because the sources of variation in race—race of parents and perhaps the social definition of race at a particular time and place—are likely to be related to the omitted variables.

Absent such so-called instrumental variables strategies, the best we are likely to be able to do with observational studies of racial discrimination is to specify the model as completely as possible. Consequently, it is critically important that the researcher understand how expected productivity is determined and obtain as rich a data set as possible. The most important variables to control for are ones that are likely to have strong effects on P and to be related to R. Because it is difficult to argue that any finite set of productivity-related factors is complete, this strategy will always yield findings vulnerable to the criticism that variables have been omitted.

Classic experimental designs employ random assignment to treatment as a way to ensure that the treatment is uncorrelated with other factors, omitted or not. If it were feasible to randomly assign people to race, then, in the context of equation (7.14), the $\gamma_2 a_1$ term would be zero, and estimating α would be a way of detecting whether a firm is discriminating on the basis of race. Similarly, random assignment of race to a pool of applicants could resolve the bias associated with sample selection. Indeed, despite their other complications, the big advantage of audit studies is their ability to manipulate race experimentally and thereby get around these problems (see Chapter 6).

Given our inability to manipulate race in observational studies, what can be done about omitted variables and sample selection bias? In the following sections, we discuss some of the strategies that have been used to address these problems.

Using an Indicator of Productivity to Address the Omitted Variables Problem

In many situations in which people are screened, such as hiring, college admission, and mortgage approval, rational nondiscriminating screeners should base their decision on how well they expect a candidate to perform if hired, admitted, or approved. In this section, we use the hiring example to show that data on actual performance can be very useful in attacking the omitted variables problem when studying discrimination.

To illustrate how such data can be employed, we consider the problem of using equation (7.14) as a basis for assessing discrimination in our hiring example. We will consider two situations: one in which the firm knows enough about the worker to make statistical discrimination irrelevant and another in which race yields nonredundant information about productivity.

Suppose that the indicator or proxy for productivity P^* is equal to

$$P^* = gP + \text{noise},$$

where the noise is unrelated to actual productivity P, X_1, X_2, and race indicator R. The researcher can then estimate $gE(P \mid X_1, R)$ from a regression of P^* on X_1 and R. In general,

$$E(P^* \mid X_1,R) = g_1 E(P \mid X_1,R) = g_1[X_1B_1 + E(X_2 \mid X_1,R)B_2 + E(Z \mid X_1,R)G_1 + E(Q \mid X_1,R)G_2].$$

In the special case in which ZG_1 and QG_2 are unrelated to R conditional on X_1 and X_2, $E(P^* \mid X_1,R)$ can be written as

$$E(P^* \mid X_1,R) = E[(P^* \mid X_1,X_2) \mid X_1,R] = g_1 [X_1(B_1^* + \gamma_1) + R\gamma_2]. \tag{7.15}$$

This is the special case in which the firm knows enough about the worker to make the information in R redundant. Consequently, the firm has nothing to gain from resorting to statistical discrimination on the basis of R. (See Chapter 4 for a discussion of statistical discrimination.) If, after conditioning on X_1 and X_2, R is still correlated with the productivity determinants Z and Q that are not observed by the firm, then equation (7.15) will not hold. Furthermore, the firm would have an incentive to resort to statistical discrimination. To see this point, note that in general

$$E(P \mid X_1,X_2,R) = X_1B_1 + X_2B_2 + E(Z \mid X_1,X_2,R)G_1 + E(Q \mid X_1,X_2,R)G_2,$$

and in this situation both R and X_1 and X_2 are useful for predicting Z and Q and thus P.

Now recall that if the firm uses only X_1 and X_2 to judge productivity, then from equation (7.14),

$$E(y \mid X_1,R) = E[E(y \mid X_1,X_2) \mid X_1,R] = \text{constant} + \alpha' a_1[X_1(B_1^* + \gamma_1) + R(\gamma_2)] + \alpha R.$$

We can identify the discrimination coefficient α by estimating equation (7.15) and then regressing y on the resulting estimate of $g_1 [X_1(B_1^* + \gamma_1) + R\gamma_2]$ and on R:

$$y = \text{constant} + \alpha'(a_1/g_1)\, g_1 [X_1(B_1^* + \gamma_1) + R\gamma_2] + \alpha R + \text{error}. \tag{7.16}$$

If equation (7.15) holds—the case in which the firm has no incentive to statistically discriminate—the coefficient on R is an unbiased estimate of α. Thus, under these circumstances, expanding the model to incorporate actual productivity permits one to solve the omitted variables problem.

If race yields nonredundant information about productivity, equation (7.15) fails. If the firm chooses to use race as a predictor and statistically discriminate, the estimate of α will be an unbiased estimate of the extent to which the hiring rate depends on R independently of the firm's belief about productivity. However, the analysis will not reveal whether the firm statistically discriminates when forming its belief about productivity.

In other circumstances, productivity data do not solve the omitted variables problem. If the firm does not statistically discriminate and the special condition of equation (7.15) fails, an estimate of α based on equation (7.16) will be biased. The reason is that the relationship between P and R and X_1 will be different from the relationship between $E(P \mid X_1,X_2)$ and R and X_1, and hiring is based on $E(P \mid X_1,X_2)$.

If the firm statistically discriminates on the basis of incorrect stereotypes about how R is related to performance conditional on X_1 and X_2, it is easy to show that the estimation of equation (7.16) will also produce biased estimates of α. The bias in α stems from the fact that actual hiring will reflect the incorrect weights placed by the firm on X_1, X_2, and R in forming its expectation of P, rather than the relationships the researcher will uncover when estimating $E(P^* \mid X_1,R)$. That is, the estimate of α will mix two forms of discrimination—the overt or subtle preference for whites given beliefs about productivity and the use of incorrect stereotypes to judge the productivity of blacks or other minority groups.

Sample selection will also influence estimates of the effect of R on productivity P if the researcher leaves X_2 out of the model. The bias in α based on the estimation of equations (7.15) and (7.16) will depend on the same considerations raised in the discussion of these two equations above.

Unfortunately, few studies have enough information about productivity to attempt to estimate an equation such as (7.16). Altonji and Blank (1999) survey a few papers examining wage differentials using a methodology that fits into the above framework, including Hellerstein and Neumark (1998, 1999), and a series of papers looking at compensation for professional athletes. (See Box 7-5 for a description of a study on racial differentials in compensation.)

A much simpler strategy is available if the researcher actually observes the firm's belief about P. In this case, one may estimate α by regressing y on R and the firm's belief. Identifying whether the firm is statistically discriminating is more difficult, particularly in the absence of data on the firm's beliefs. Progress can be made using variants of the approach of Altonji and Pierret (2001), although strong assumptions are required.

BOX 7-5
Racial Differentials in Compensation: The Case of Professional Athletes

A number of studies of compensation of professional athletes closely follow the line of analysis outlined here. Kahn and Sherer's (1988) study of professional basketball is a good example. They hypothesize that the value of a player depends on his marginal revenue product. They assume that marginal revenue product (the effect of a player on the team's revenues) depends on the player's actual performance and on the racial preferences of the team's fans. They hypothesize that performance is a function of a set of characteristics of the player, including total seasons played, average minutes per game, career free throw percentage, career per-game steals, and so on. They estimate a model relating compensation to these characteristics and race, finding that whites are paid about 20 percent more than blacks with the same characteristics. The study goes on to attempt to determine whether the race gap is due to customer discrimination. Kahn and Sherer find that replacing one black player with a white player with the same performance statistics raises home attendance by 8,000 to 13,000 fans per season. They suggest that part of the salary differential may reflect the response of owners to customer discrimination.

The joint analysis of productivity and hiring is a step further down the road toward working with a complete model of the process through which hiring and discrimination operate. We have demonstrated how a more complete model can be used to draw inferences when omitted variables are present, which is in the spirit of Fisher's suggestion that theories be made more elaborate. Furthermore, an important advantage of working with a joint model of productivity and hiring is that some of the basic assumptions underlying the estimation of α become testable. For example, the model implies that within a group, hiring decisions are based on expected productivity. This assumption is testable. We want to emphasize that, even here, strong assumptions about how the hiring process operates must be made to infer the effect of race on hiring decisions.

In Annex 7-1, we consider how productivity data can be used to detect adverse impact discrimination, which we define as adopting hiring criteria in ways that are not justified by productivity considerations and that are harmful to a minority group.

Matching and Propensity Score Methods

Matching methods provide an alternative to multivariate linear regression as a way to control for variables that are likely to matter for an outcome in observational studies. Matching consists of comparing outcomes of two paired individuals (or groups) who are comparable on relevant observed attributes except for race. Matching attempts to mimic the experimental setting in the same way as paired testing. To the extent that (1) the observed factors capture the relevant variables affecting the outcome and (2) the comparability is close, racial differences in the outcome variable in a matching study can be attributed to discrimination. Matching has been the subject of considerable research, and relatively sophisticated methods, such as propensity score matching, have been developed.

The objective in matching is to construct matched sets or strata using relevant nonracial covariates that are available. Analogous to overfitting in specifying a multiple regression, the analyst doing the matching must make the trade-off between matching on too few variables with the result of poor comparability within matched sets and matching on too many variables with the result of poor statistical power and problems with interpretation (i.e., overmatching). A common way to manage this trade-off is to combine matching on a small number of variables that are proven to have large effects together with matching on propensity scores (described below) derived from a larger set of additional variables thought to be relevant.

Propensity score matching addresses the problem that, as the number of covariates increases, it becomes increasingly difficult to find matched sets with similar values of the covariates. Even if each covariate is binary, there will be 2^p possible sets of covariate values leading to very fine-grained strata for large p. The propensity score is a device for constructing matched sets or strata when there are many covariates. It is typically estimated by fitting a logistic regression to minority versus nonminority group membership using the covariates as the explanatory variables. Subjects with similar propensity scores are grouped into the same strata to create matched sets.

As compared with multiple regression, matching methods reduce the risk of imposing an inappropriate functional form on the relationship between the outcome y and the observed covariates. Multiple regression models use all the data. In matching, on the other hand, each minority-race individual is typically matched to one or more nonminority individuals. The pool of unmatched nonminority individuals is not used in a matched analysis. Matching is most effective when the minority group is very small as compared with the nonminority group. In this situation, the loss in precision from discarding the nonmatched members of the nonminority group is low. Choosing between matching and regression methods often involves weighing the trade-off between reduced sample size from matching and the functional-form assumptions needed for regression. See Rosenbaum (2002) for an excellent review of these methods and a discussion of the advantages and disadvantages of matching versus multiple regression in various situations. As noted above, matching techniques are beginning to be used as an alternative to multiple regression in statistical decompositions of racial differences. However, these methods do not help with the key problems of omitted variables bias or sample selection bias because matching is performed on the basis of observed variables only.[8]

In the same spirit as matching, stratification on relevant variable(s) can also be used to achieve some measures of control on nonracial factors. Stratification approaches include (1) pre- and poststratification of the data for analysis purposes and (2) adjustment of strata to a standard population. Because stratification methods are widely used in the epidemiology literature but not in the statistical discrimination literature, we simply refer the reader to Sarndal et al. (1992) for further information.

Panel Data Methods

Another strand of research using observational data has proceeded by exploiting features that become available with longitudinal data. People do not change race, but one can learn about changes in the consequences of race by following individuals over time. If unobserved characteristics are relatively stable, but outcomes and factors that are related to discrimination do change over time, then the unobserved factors can be partialed out of analyses comparing racial groups over time. (For an overview of the analysis of panel data, see Hsiao, 1986.) In the labor market context, following individuals over time enables one also to examine the differences across regions, industries, sectors (private versus government), and occupation in racial differences in outcomes. The assumption that unobserved factors are not changing over time is a strong one, however, and that assumption typically cannot be approximated well enough to be usable unless the time frame is relatively short. Furthermore, people do not switch regions, change industries, and so on at random, which suggests that longitudinal designs comparing the consequences of changes for whites and minorities are subject to selection bias.

---

[8]Rosenbaum (2002) discusses methods of examining the sensitivity of results to quantitative assumptions about the correlation between the variable of interest (race in the present case) and omitted factors that may be useful for producing bounded estimates of the discrimination coefficient α (see also Manski, 1995). Manski (2003) discusses other approaches to construction of bounds under very weak assumptions about omitted variables.

Altonji and Pierret (2001) provide an example of how, at the cost of some very strong assumptions, longitudinal data can be used to draw inferences about discrimination (see Box 7-6).

Natural Experiments

Another approach to addressing the problem of omitted variables and limited understanding of how a nondiscriminating firm would make decisions is to exploit so-called natural experiments. As discussed above, using the experimental approach in a controlled setting makes it easy for an experimenter to directly ascertain the effects of explanatory variables on some outcome of interest. If the experimenter could force some firms to be

BOX 7-6
Use of Longitudinal Data to Draw Inferences About Discrimination

Altonji and Pierret (2001) explore the implications of a hypothesis they refer to as "Employer learning with statistical discrimination," using the National Longitudinal Survey of Youth, 1979. If profit-maximizing firms have limited information about the general productivity of new workers, they may choose to use easily observable characteristics, such as years of education or race, in judging workers, even if doing so means violating the law in the case of race. The theory put forth by Altonji and Pierret implies that, as firms acquire more information about a worker, pay may become more dependent on productivity and less dependent on easily observable characteristics or credentials. One implication of their model is that, (1) if blacks and whites differ in labor market productivity because of difficult-to-observe premarket factors such as school quality and (2) if employers statistically discriminate against blacks, this situation will lead to a race gap in initial wages, although it cannot easily explain race differences in wage growth. On the other hand, if firms do not statistically discriminate on the basis of race, a wage gap will open up as firms directly observe productivity. Altonji and Pierret's empirical results call into question the statistical discrimination explanation of wage differences because they observe that the race gap in wages is small at labor market entry and grows with experience in the labor market. The authors point out many caveats to their study, including the possibility that differences between blacks and whites in access to training and promotion opportunities, perhaps because of other, more overt, forms of discrimination, may explain some of their findings. With data on productivity and training as well as wages, the analysis would be strengthened.

nondiscriminators, their behavior could be compared with that of a control group of firms, and discrimination could be measured as the difference. When an experimental design is not practical, researchers can use natural experiments to observe the natural variations that occur both before and after a specified time period during which an intervention is introduced. Thus, the researcher observes some exogenous intervention, such as a policy change that affects procedures governing hiring or college admission. Instead of random assignment, treatment and comparison groups are defined, and naturally occurring events are used for comparisons.

Social scientists have used a "differences-in-differences" approach (i.e., the racial difference in some outcome of interest both before and after an intervention) to test the effects of changes occurring at some specified time period that affect some firms or other actors but not others (see, e.g., Card and Krueger, 1994; Tyler et al., 1998). In the language of causal modeling, the policy change is a formal manipulation, which is applied to some actors (e.g., firms in a particular industry or state) but not others. (In some studies, the policy change affects all actors and the comparison is done before and after the change.) The pre-policy-change data are used to estimate the counterfactual condition of what would have happened had the policy change not occurred. Such designs are also sometimes termed quasi-experiments. Because there is some degree of control, the assumptions made for natural experiments to support a causal inference need not be as strong as those required for uncontrolled observational studies; however, natural experiments fall short of randomized controlled experiments. (For more detail on natural experimental designs, see Campbell and Stanley, 1963; Meyer, 1995; and Shadish et al., 2002.)

The key idea in the context of discrimination is to find settings in which there is an exogenous source of variation in the weight that firms, schools, lenders, and other actors could place on R that can be plausibly thought of as being independent of the unobservable factors. In the hiring case, the idea is to contrast hiring policy before and after a change in procedure that restricts or eliminates the extent to which the firm can use race in hiring decisions.

Examples of Natural Experiments to Study Discrimination

Below we provide several examples of natural experiments that were used to study racial discrimination in the labor market, education, and health care domains.

Labor market. Holzer and Ludwig (2003) discuss the empirical research using natural experiments in the labor market domain to test the effectiveness of antidiscrimination laws (e.g., Freeman, 1973; Heckman and Payner,

1989) and the effects of a policy change on different racial and ethnic groups (e.g., Chay, 1998; Neumark and Stock, 2001). The basic approach is to determine whether the law has had the result of forcing α to be zero, or at least closer to zero than was the case before the law. By comparing employment rates before and after the change in the law, one can draw inferences about the extent of reduction in discrimination after the change.

Another source of variation in employment policy to examine could be changes at the firm level within an industry that are not necessarily mandated by changes in the law. Goldin and Rouse (2000), for example, focused on an interesting setting for an attempt to measure discrimination in hiring—a professional orchestra. Even though they considered discrimination against women, the methodology they used applies to other groups as well. In the 1970s and 1980s, many orchestras began to hide the auditioning musician from the jury by using a screen or other device. The representation of women in orchestras, especially among new hires, increased dramatically over this period, and the question investigated was the extent to which this increase was attributable to the change in audition practices. Goldin and Rouse obtained data for a number of years on the auditions for a set of nine orchestras; some of these auditions took place with the screen, while others did not.

Basically, Goldin and Rouse used regression models to estimate whether an individual advanced from one round of auditions to the next and whether an individual was hired in the final round as a function of three things: (1) type of audition (blind versus not blind), (2) the interaction between gender and type of audition, and (3) controls for characteristics of the individual and the audition. By construction the weight on gender in the blind audition was 0, because the gender was unknown to those judging the musicians. Because the researchers had multiple observations of the orchestras, they were able to include orchestra-specific constants to control for unobserved differences among orchestras that might have been associated with adoption of a screen. They were also able to include person-specific constants to control for unobserved differences in the quality of the musicians to guard against the possibility that the use of a screen influences the relative quality of the men and women who audition, which made the results weaker.

Overall, Goldin and Rouse found that women were much more likely to advance and be hired when auditions were blind and concluded that the introduction of a screen led to reduced discrimination against women. Their work provides an example of a case in which a change in policy made discrimination at a key stage of the hiring process much more difficult than it had previously been. Moreover, the increase in the rate of hiring of women after the change demonstrated that discrimination existed prior to the change. The obvious limitation of this work is that it is dangerous to draw broad conclusions about discrimination in hiring from the orchestra case.

Yet another source of variation in employment policy to examine could be changes in wages in an industry in response to competitive pressure. The idea is that prejudiced firms may indulge biases when economic rents (excess profits) are available. Competitive pressure may reduce the rents, forcing firms to either reduce discrimination (hire the best people) or go out of business. This situation would lead to a reduction in α, the weight placed on race.

Black and Strahan (2001) provide one of the cleanest of these studies, and although they focused on gender discrimination, the idea can be applied elsewhere. They exploited the fact that regulations constraining entry by banks into new markets were relaxed beginning in the mid-1970s. Using data from the mid-1970s through 1997, they found that following deregulation the average wages of bank employees declined relative to the wages of nonbank employees. The authors used multivariate regression models to implement a triple-differencing strategy to distinguish the effect of deregulation from fixed characteristics of states and wage and employment trends at the state level that happen to be correlated with deregulation. The strategy amounts to taking the difference between the growth in wages of bank and nonbank employees in states that undergo deregulation at a certain point in time and comparing it with the corresponding difference in wage growth rates for bank and nonbank employees in states that did not undergo deregulation at that point in time. Black and Strahan show that deregulation led to a decline in the gap between the wages of men and women for two reasons: First, women moved into higher-skill occupations; second, the wages of men fell more than the wages of women in a given occupation. This evidence is consistent with some models of gender discrimination.

Health care. Chay and Greenstone (2000) examined trends in black–white infant health outcomes between 1955 and 1975. The authors fit simple trend-break regression models to vital statistics data for blacks and whites in rural and urban areas of different states. They used a time trend variable to measure the average trend in infant mortality rates across states (1955–1965) and an indicator variable to measure the change in the infant mortality trend after 1965. They controlled for differences across states (by race and rural versus urban area) that might be correlated with infant mortality.

The regression results showed a significant trend break in health outcomes for black and white infants after 1965, although improvements were more pronounced for blacks. The authors note that before 1965 black infant mortality rates were high relative to whites. Between 1965 and 1975, however, there was evidence of a sharp decline in black infant mortality rates and convergence of these rates after 1965 (particularly in the rural South). Chay and Greenstone suggest that the implementation of two federal interventions—Title VI of the 1964 Civil Rights Act (prohibiting discrimination and segregation in access to care) and the Maternal and Child Health Services Program under Title V of the 1935 Social Security Act[9]—could explain the convergence of black–white infant mortality rates after 1965.

Because the trend-break patterns showed similar improvements for whites across all regions after 1965, it is possible that other causal factors along with race might explain the post-1965 changes. The authors also report a strong correlation between "differential convergence in infant mortality rates" and "differential convergence in black–white hospitalization rates across states" (2000:330). Thus, the federal interventions, and possibly other factors, played an important role in the changes in relative infant mortality rates.

Education. Holzer and Ludwig (2003) provide some examples of natural experiments in the education domain that can be used in research to examine the effect of racial differences in educational inputs on relative outcomes. One type of natural experiment in the education domain looks at discriminatory educational policies and practices and assesses their effects on education outcomes. Examples are studies looking at the adverse effects of "separate but equal" laws on the educational attainment of blacks prior to the ruling in Brown v. Board of Education (Boozer et al., 1992; Donohue et al., 2002; Margo, 1990) and studies of the effects of school desegregation orders implemented after the Brown ruling (Guryan, 2001). Such experiments can be useful for measuring discriminatory practices in education but are difficult to apply in this domain. Holzer and Ludwig (2003:1167) offer this perspective:

> Evaluating how these natural experiments change the allocation of educational inputs across or within schools may help highlight the degree to which racial discrimination affected educational decisions in the past. One limitation with this approach is that social scientists are limited to either detecting discrimination within a given jurisdiction retrospectively rather than prospectively, or must extrapolate from evidence of past discrimination in one jurisdiction to other areas where policy makers seek guidance on future enforcement or policy actions.

[9]The Maternal and Infant Care component, expanded in 1963 and 1965, was established to improve the health of mothers and infants in low-income and rural families.

Another type of natural experiment focuses on the general relationship between educational inputs and outputs. For instance, one such experiment might examine the effects of a policy change in tracking or ability grouping on student outcomes. Differences in student outcomes within one school before and after the policy change could be compared with outcomes in another school that did not experience a policy change to determine whether discrimination played a role (see Holzer and Ludwig, 2003, for further details). Holzer and Ludwig conclude that natural experiments are valuable tools for determining whether observed racial differences in inputs constitute racial discrimination and for measuring the effects of such differences.

Limitations of Natural Experiments

In the context of the study of discrimination, as well as in other arenas, natural experiments have limitations. First, the change under study may be endogenous. That is, it may be a reaction to particular circumstances that warranted a policy change or intervention. As a result, one may not be able to generalize from the results of a study to estimate the average amount of discrimination prior to the change. For example, suppose one is trying to measure discrimination by comparing hiring rates in a particular firm before and after an intervention by the Equal Employment Opportunity Commission (EEOC) with those of firms in the same industry around the same time period. Assuming that the EEOC responds to the most serious cases, the estimated effect would tend to overstate the amount of discrimination in the industry at large prior to the intervention.

Second, the effects of policy interventions may spill over into the control groups used in the study. For example, the effects of heightened EEOC activity involving a particular set of employers in a given industry might influence the behavior of other firms and industries even though they have not been targeted. This phenomenon would reduce estimates of the effect of EEOC enforcement based on a "differences-in-differences" design.

Third, differences in trends in other factors that affect outcomes cannot always be addressed adequately even in differences-in-differences designs, particularly when the policy intervention takes place over a period of time, as is the case with civil rights policy.

Fourth, a change in one domain, such as school desegregation orders, may be accompanied by changes in another domain, such as housing, or by a change in attitudes. Consequently, it may be difficult to use a change in policy in one domain to identify the amount of discrimination in that domain prior to the change.

Fifth, only in rare circumstances (such as the Goldin and Rouse orchestra study) can one be sure that the change in policy under study has eliminated a role for discrimination in the decision under study. In most cases, the best one can hope for is that a comparison of groups affected by the change in policy will identify the reduction in discrimination induced by the policy (the change in α), rather than the level of discrimination that existed prior to the change.

Sixth, in some cases, changes in policy that lead to positive effects in one dimension may induce negative effects in others. A major concern in the literature on the effects of antidiscrimination policy in the labor market, for example, is that positive effects on wage rates for blacks have been offset in part by negative effects on employment (see Altonji and Blank, 1999, for discussion and references).

Finally, natural variation in the data may be insufficient to identify the effects of interest or may be correlated with other, unmeasured factors that may bias the results. (See Holzer and Ludwig, 2003, on the use of natural experiments to study discrimination; see Shadish et al., 2002, and Meyer, 1995, for a general discussion of the strengths and weaknesses of these designs.)

Summary of Possible Solutions to Problems of Using Statistical Models to Infer Discrimination

It should be obvious that more accurate and complete data collection efforts are critical to reducing the key problem of omitted variables bias. Of course, the data needed must pertain to the particular domain of analysis. Data on performance (e.g., productivity in the hiring context, default rates in the lending context) and detailed knowledge of how an outcome depends on performance can solve the problem of omitted variables bias in some cases. However, situations in which the researcher will possess the data and detailed knowledge needed to support specification of an appropriate model are relatively scarce, at least in the labor market setting.

Matching and propensity score methods are useful as a means of relaxing assumptions about the functional form relating the variables X_1 and X_2 to productivity and to hiring decisions. However, they do not solve the omitted variables bias problem.

Panel data are useful as a way of identifying differences in the amount of discrimination across types of institutions, regions, or time. However, this approach requires the assumption that time-varying unobserved characteristics of the individual are not related to mobility, which is a strong assumption.

Natural experiments in which a legal change or some other change forces a reduction in or the complete elimination of discrimination for some groups provides leverage in assessing the importance of discrimination prior to the change and for groups not affected by the change.

ADDITIONAL ISSUES

Thus far we have discussed prospects and problems for measuring discriminatory treatment of persons who are identical except for race. In the context of our hiring example, the parameter α measures discriminatory treatment of blacks and whites with the same values of X_1 and X_2, the variables known to a firm to determine productivity. The model developed above, however, also sheds light on other discriminatory processes.

Effects of Past Labor Market Discrimination on Factors in Hiring

Discrimination by a firm or elsewhere in the labor market may influence some of the elements of X_1 and X_2, the (nonracial) characteristics used by the firm to make hiring decisions. For example, some studies of hiring are based on whether the person was referred by an existing employee (Fernandez and Weinberg, 1997; Fernandez et al., 2000; Petersen et al., 2000). Current labor market discrimination against minorities by the firm will lead to a discrepancy in the probability that minority applicants will know people who work in the firm because personal networks tend to run along racial lines. Even if the use of referrals in hiring is justified by productivity considerations, the total effect of current discrimination will be understated if one holds constant whether a person was referred to the firm. In particular, the parameter α will understate current discrimination.

Alternatively, suppose that in the past a firm discriminated against disadvantaged racial groups but no longer does. Continuing with the referral example, again suppose that the firm makes use of referrals in hiring decisions. If the researcher is simply interested in whether the firm treats applicants with a given set of characteristics differently at the present time, and if the researcher observes all the variables the firm uses to assess productivity (there are no X_2 variables that the researcher does not observe), the researcher will draw the correct conclusion of no such differential treatment. In this case, α will be zero. However, if the researcher wants to know the total effect of both past and current discrimination on the part of the firm on the racial composition of current hires, it is incorrect to take as given whether employees were referred. The reason is that past discrimination led disadvantaged racial groups to be underrepresented among the pool of potential referrers, thus reducing the chances of attracting disadvantaged racial groups through referrals. To measure the effect of both past and current discrimination on current outcomes in this dynamic context, the researcher must model the effect of past discrimination on current X variables.

To give another example, it is standard practice for many types of jobs—and in many situations defensible from a productivity standpoint—

to consider past work experience when trying to predict productivity. However, past experience will be influenced by discrimination in the labor market. Consequently, the coefficient on R will provide an estimate of the effect of discrimination on the firm's behavior, given X_1 and X_2. But because discrimination in the labor market leads to a racial gap in the experience-related components of X_1 and X_2, the coefficient on R will understate the total effect of all discrimination that has taken place in the labor market.

Furthermore, discrimination in the labor market may influence the choices of X_1 and X_2 that people make before they enter the labor market. For example, if African Americans know they are discriminated against for white-collar positions and college has little value in the blue-collar world, they will have less incentive to pursue a college education. If one uses a model such as equation (7.13) to measure labor market discrimination against blacks in white-collar positions holding education (one of the X_1 variables) constant, one will underestimate the total effect of discrimination on the racial composition of such jobs. Similarly, if firms develop a reputation of having a hostile work environment for racially disadvantaged groups and if such applicants avoid seeking employment at those firms, a model such as equation (7.13) will underestimate the total effect of discrimination in hiring decisions. This will be the case even if the researcher observes all of the variables used by firms to choose employees. Developing measures of the discrimination that results from a process such as that described above is extremely challenging because of the much longer timeline and more complex environment that must be accounted for to reach statistically valid "all else being equal" conclusions. We address these issues in more detail in Chapter 11.

Effects of Discrimination in Other Domains

To measure the total effect of discrimination in society on a particular outcome, such as the odds of getting hired, one needs to measure the effects of discrimination in other domains on elements of X_1 and X_2 that are determined outside of the labor market (see Chapter 11). In our example of hiring, if there is racial discrimination in "pre-market factors" (Neal and Johnson, 1996), such as education, that are related to labor market success, discrimination in the educational sphere will also affect labor market success indirectly. Thus controlling for education in a hiring equation is reasonable in assessing whether a particular employer is discriminating.

However, if there is racial discrimination in the educational domain, controlling education will understate the total effect of all racial discrimination in analyses of labor market discrimination alone. Developing and validating statistical models of these broader processes is one way to gain insight into the presence or absence of discrimination in these other areas.

One's choice of control variables is influenced by whether one is trying to measure discrimination in a specific domain or the cumulative impact of discrimination.

Other Discriminatory Effects on the Productivity Equation

Another issue concerns whether discrimination on the part of customers, coworkers, or suppliers leads characteristics of the worker, *including race*, to enter the productivity equation (7.9).[10] Consider Becker's (1957) theory of customer discrimination, and consider sales positions. Suppose that white customers prefer to buy from white salespeople, and black or Latino customers are indifferent. In such a world, P is influenced by the match between the race (or ethnicity) of the job candidate and the racial composition of the customers. R will not appear directly in the equation for productivity, but the interaction between R and the racial composition of the customer base will. If the firm obeys the law, it will not apply the interaction variable in making decisions about hiring, and the interaction variable will not enter significantly into hiring decisions. (The interaction will show up in a productivity regression.) If the firm disobeys the law, the interaction term will influence hiring and show up in a hiring regression. One will then conclude correctly that firms discriminate for or against black or Latino salespeople as a function of the customer base. If one excludes the interaction term but adds R to the hiring equation, one will likely find evidence that the firm discriminates against minorities if most of the markets for which the firm is hiring happen to be heavily white. But one will not detect the fact that the nature of the discrimination is related to the match between customers and the sales agent. If there are data that can be used to estimate the effect of the interaction between race and customer composition on productivity, one can see whether hiring decisions appear to reflect such considerations. A number of studies of professional sports take this approach (see Altonji and Blank, 1999, and Kahn, 1991, for examples).

A somewhat different example involves the possibility that discrimination in social institutions that are extraneous to the firm or the labor market influences the form of the productivity equation. Consider again a marketing position. Suppose that social connections play a critical role in marketing. In such a world, sales productivity may well depend on club memberships, where one lives, the schools one attended, and the like. Variables measuring such social connections belong in X_1 and X_2. R may have no relationship to productivity or to hiring decisions if one conditions on

[10]Throughout this section we are defining productivity to be the effect of an employee on the profitability of a firm; we are excluding societal objectives.

these variables. Now suppose that societal discrimination (including housing discrimination) influences social connections. In this case, discrimination outside the labor market will lead to a race gap in some elements of X_1 or X_2 or both, as well as in hiring, even though the variable R will not have an independent effect on hiring conditional on X_1 and X_2. Finally, note that the recruiting strategies chosen by the firm are likely to influence the importance of social networks. Strategies that place more emphasis on personal contacts and less on advertising may not be race neutral. A discriminating firm may consciously choose a recruiting strategy in which social networks are important and then exclude minorities who lack them.[11] It will be difficult to determine whether the firm's recruitment strategy is really the profit-maximizing one or in fact is shaped at least in part by the goal of discrimination.

SUMMARY, CONCLUSIONS, AND RECOMMENDATIONS

The main purpose of this chapter has been to review the strengths and limitations of various approaches to dealing with the challenges of measuring discrimination with statistical methods using observational data. This review leads to several conclusions.

Our first conclusion relates to the uses and limitations of statistical decomposition of gaps in outcomes among racial groups:

> **Conclusion:** *The statistical decomposition of racial gaps in social outcomes using multivariate regression and related techniques is a valuable tool for understanding the sources of racial differences. However, such decompositions using data sets with limited numbers of explanatory variables, such as the Current Population Survey or the decennial census, do not accurately measure the portion of those differences that is due to current discrimination. Matching and related techniques provide a useful alternative to race gap decompositions based on multivariate regression in some circumstances.*

More generally, we will often be hampered in our ability to infer discriminatory behavior on the basis of regression decompositions because we can never be sure we have included all the relevant controls in the model. We must be able to control for the relevant variables well enough to approximate closely the hypothetical counterfactual in which only race has been changed.

---

[11]Rees and Shultz (1970) give an excellent example of a firm that used its recruitment methods to discriminate. They report that a Chicago-area steel mill advertised only in Polish-language newspapers in an effort to avoid African Americans.

Our second conclusion follows naturally from the first:

Conclusion: *Nationally representative data sets containing rich measures of the variables that are the most important determinants of social and economic outcomes—such as education, labor market success, and health status—can help in estimating and understanding the sources of racial differences in outcomes. Panel data may be particularly important and useful (see Chapter 11).*

Not only must statistical models for estimating discrimination use appropriate data and methods, but they must also be based on as thorough as possible an understanding of the processes that underlie the behavior being studied. Otherwise, the models are likely to require strong assumptions that cannot be justified. More generally, the properties of the model used for analysis are crucial in assessing claims of statistical "proof" of discrimination. Researchers must provide sufficient information on their model to enable others to understand and make a judgment about whether the assumptions underlying the model have been met.

Conclusion: *The use of statistical models, such as multiple regressions, to draw valid inferences about discriminatory behavior requires appropriate data and methods, coupled with a sufficient understanding of the process being studied to justify the necessary assumptions.*

The specific model we developed in the context of the decision to hire a worker illustrates the role played by assumptions and theory in drawing causal influences based on observational data. It also sheds light on how omitted variables and sample selection biases affect our ability to draw conclusions about discrimination and helps make clear what forms of discrimination are measured and what forms are not.

Data on performance relevant to a particular domain, such as productivity in the labor market context or academic success in the educational arena, are extremely valuable in dealing with the problem of omitted variables bias, in permitting the testing of key assumptions of a statistical model, and in studying adverse impact discrimination (see Annex 7-1 below). Natural experiments, although they have limitations, provide another way to address the problems of omitted variables bias and limited knowledge of the decision processes of particular actors.

Conclusion: *We see an important role for focused studies that target particular settings (e.g., a firm or a school), whereby it is possible to learn a great deal about how decisions at each stage in a process are made and to collect most of the information on which decisions are* *based. With such knowledge and data, it becomes much easier to specify an appropriate statistical model with which to estimate racial discrimination.*

Conclusion: *Despite limitations, natural experiments—in which a legal change or some other change forces a reduction in or the complete elimination of discrimination against some groups—can provide useful data for measuring discrimination prior to the change and for groups not affected by the change.*

Recommendation 7.1. Public and private funding agencies should support focused studies of decision processes, such as the behavior of firms in hiring, training, and promoting employees. The results of such studies can guide the development of improved models and data for statistical analysis of differential outcomes for racial and ethnic groups in employment and other areas.

Recommendation 7.2. Public agencies should assist in the evaluation of natural experiments by collecting data that can be used to evaluate the effect of antidiscrimination policy changes on groups covered by the changes, as well as groups not covered.

ANNEX 7-1: DETECTING ADVERSE IMPACT DISCRIMINATION

We discuss here ways to detect adverse impact discrimination; that is, discrimination by using factors that correlate with race. A firm may not use race directly, but it may weight variables in hiring decisions in a way that is not proportionate to their influence on productivity. For example, suppose the firm uses

$$P^f = X_1(B^*_{1f}) + X_2(B^*_{2f})$$

as its productivity rating rather than the correct index

$$E(P \mid X_1,X_2) = X_1(B^*_1) + X_2(B^*_2)$$

and hires accordingly. In this case, y will be determined by

$$y = \text{constant} + \alpha' a_1\, [X_1(B^*_{1f}) + X_2(B^*_{2f})] + \alpha R + u, \qquad \text{(A7.1)}$$

where α' is $(1 - \alpha)$ as before. It is quite possible for α to be 0 even though the firm's hiring rule has an adverse impact on R that is not justified by productivity considerations. That is, α can be zero even though R is systematically related to the difference between the index P^f and the unbiased productivity index $E(P \mid X_1,X_2)$. We define this as adverse impact discrimination. The legal requirement that firms validate hiring criteria having an adverse impact on protected classes of workers is designed to prevent this form of discrimination.

In general, it will be difficult to detect that the firm is behaving in accordance with equation (A7.1) without information on P. Suppose, however, that the researcher has an unbiased indicator P^* of P as well as data on X_1 but not X_2. Then the researcher can estimate the coefficients θ_1 of the conditional expectation

$$E(P^* \mid X_1) = X_1\theta_1.$$

If firms are hiring on the basis of expected productivity given X_1 and X_2, then $E(y \mid X_1) = E(y \mid X_1\theta_1)$. Consequently, one can test the null hypothesis that firms are hiring on the basis of expected productivity given X_1 and X_2 by testing the restriction that

$$E(y \mid X_1) = E(y \mid X_1\theta_1).$$

One can test this restriction by regressing y on $X_1\theta_1$ and X_1 (with one element of X_1 excluded because of collinearity) and testing the null hypothesis that the elements of X_1 have no effect on y, holding $X_1\theta_1$ constant. From a regression of $E(y \mid X_1)$ on R and $X_1\theta_1$, one can estimate the race gap for workers with a given value of X_1 that is due to the firm's policy. Without special assumptions, however, one cannot estimate the effect on group R of the firm's misuse of X_1 and X_2 without having data on both variables. Unfortunately, even a noisy indicator of productivity is unavailable in most of the data sets used to study racial differences.

8

Attitudinal and Behavioral Indicators of Discrimination

Thus far we have discussed experimental and nonexperimental approaches to drawing causal inferences about the presence and effects of racial discrimination in one or more domains. As discussed in the previous chapter, a primary challenge to researchers attempting to draw such inferences is the inability to identify and control for all the characteristics that might affect an outcome. In this chapter, we consider the challenges of collecting direct evidence of the incidence, causes, and consequences of discrimination from sample surveys, governmental administrative data, nongovernmental data, and in-depth interviews. We also review in some detail the scale measures of racial attitudes used in surveys.

Survey and administrative records data do not provide direct observations of actual discrimination, but they can measure reported experiences, perceptions, and attitudes that involve discrimination. Used in addition to statistical methods for drawing causal inferences, these data can help researchers better understand the nature and likely occurrence of racial discrimination.

Several parts of this chapter are drawn from a paper commissioned by the panel by Thomas Smith, director of the General Social Survey (GSS), on various approaches to measuring racial discrimination and their strengths and weaknesses (Smith, 2002). The approaches he assesses include administrative counts of discriminatory incidents reported to governmental and nongovernmental authorities, in-depth interviews about past experiences with discrimination, and survey-based studies in which representative samples of randomly selected respondents report on their experiences with discrimination and prejudice. Some approaches (e.g., large-scale surveys) assess aggregate knowledge about the incidence of discrimination, while other approaches (e.g., information obtained from in-depth interviews) assess knowledge of such occurrences in a particular instance.

THE CHALLENGE OF DIRECT MEASUREMENT OF DISCRIMINATION

Prior to 1964, many forms of racial discrimination were legal (see also Chapter 3). Although a few states had passed civil rights laws, there was no national prohibition on discriminatory treatment. Many Americans were regularly denied access to jobs, housing, accommodations, and services on the basis of their race or ethnicity. Under these circumstances, the incidence, causes, and consequences of discrimination could be assessed by surveying employers, landlords, and business establishments about their racial policies or by surveying minority group members about the frequency and circumstances under which they were denied access to jobs, housing, or accommodations because of their race.[1]

The civil rights legislation passed in 1964 made open discrimination on the basis of race or ethnicity illegal, and perpetrators could be prosecuted under both criminal and civil law. In succeeding years, overt discrimination was prohibited progressively in various markets in the United States—first in markets for labor, goods, and services (in 1964); later in housing markets (in 1968); and finally in mortgage lending (in 1974, 1975, and 1977). Overtly denying nonwhites a job or apartment because of their race would invite prosecution and, if admitted in court, would virtually guarantee conviction.

Although readily observable acts of discrimination declined, the persistence of high levels of residential segregation along racial lines and large racial gaps with respect to income, wealth, and other societal outcomes indicate the continued existence of racial discrimination, at least at some level. Rather than being open and readily observable, however, discrimination was more often subtle in nature, assuming new forms that are not as easily identified but may be damaging nonetheless.

Under these circumstances, assessing the extent of racial discrimination directly from observational studies, such as asking black Americans whether they have been denied housing or employment because of their race, can be complicated. Those surveyed may suspect that they have been victims of discrimination, but ambiguity in such a situation can make proving discrimination—or sometimes even recognizing it—difficult. On the one hand, the true incidence of discrimination may be underestimated because a per-

[1]Although in some cases, social pressure may have impaired the accuracy of such reports.

son may not realize that it has occurred or be uncertain about whether it was actually present. On the other hand, the true incidence of discrimination may be overstated because, in an ambiguous situation, respondents may falsely attribute the denial of work or housing to discrimination that is in fact due to some other reason, such as qualifications, timing, or even chance. Likewise, asking white Americans whether they intend to discriminate or whether they support discriminatory policies is unlikely to provide a good indication of the prevalence of racial discrimination in American society (e.g., see Bonilla-Silva and Forman, 2000). Respondents to social surveys are understandably reluctant to admit to socially sanctioned behavior, particularly if it is illegal (Sudman and Bradburn, 1982). Thus, whites will tend to underreport discriminatory attitudes and behaviors and be unwilling to admit to perceptions or feelings that might appear to be prejudiced (a phenomenon known as social desirability bias).

Much of the information on racial attitudes and perceptions of discrimination comes from sample surveys and other observational studies (as distinct from experimental studies). It is therefore important to be cognizant of what such data sources can and cannot do. Survey measures capture reported experiences, perceptions, or attitudes about discrimination. These measures may provide some indication that discrimination is occurring, as well as what its causes and consequences may be. Nonetheless, evidence from surveys and other observational studies is not statistically valid in terms of causality. In such studies, it is not discrimination per se that is being directly measured but reports of experiences of discrimination or discriminatory attitudes.

In Chapter 7, we discussed the difficulty of using observational data to assess the extent to which racial differences in outcomes are the result of discrimination, as opposed to some other observed or unobserved variable that is correlated with race. Unknown characteristics or process-related differences in treatments (e.g., processes occurring before a reported incident) can bias observed effects. Because only a small set of characteristics is often available in any data set, other unobserved characteristics that may affect the outcome of interest are omitted, biasing the observed effect of racial discrimination. Thus, we again see how lack of experimental control can hamper the ability to draw causal inferences from observational data.

To the extent that members of disadvantaged racial groups report being discriminated against and whites admit to racist attitudes and discriminatory behaviors, these data likely represent lower-bound estimates of the actual occurrence of discrimination in society, although evidence on this issue is far from clear. The only "validation" of reported discrimination comes from credible descriptions of discriminatory incidents and from instances in which self-reports, matched-pair studies, and counts of complaints yield similar results, particularly similar intergroup differentials (Smith,

2002). Indeed, direct measures of experiences and perceptions of discrimination are probably best used to support valid findings from other kinds of studies to estimate the contribution of discrimination to observed disparities in outcomes among racial groups. Longitudinal data in particular can provide useful information about process-related differences across racial groups, such as changes in perceptions and experiences (as well as outcomes) over time.[2] Thus, despite our caveats, we conclude that it is informative to consider directly reports of levels and changes in discrimination and prejudice.

SOURCES OF OBSERVATIONAL DATA

Probably the most extensive data on experiences and perceptions of racial discrimination come from survey research. We therefore begin this review of sources of observational data on discrimination by discussing in detail the content, uses, advantages, and limitations of surveys of interpersonal relations and racial discrimination. We then examine more briefly three additional data sources: governmental administrative data, nongovernmental data, and in-depth interviews.

Surveys of Interpersonal Relations and Racial Discrimination

Design and Strengths

Surveys of interpersonal relations use interviews or questionnaires to collect detailed information about respondents' perceptions of and experiences with discrimination. Researchers use this information to better understand the causes and consequences of various forms of discrimination and to investigate relationships among racial attitudes and beliefs, on the one hand, and perceptions and experiences of discrimination on the other. The surveys are typically administered on a regular basis, either cross-sectionally or longitudinally, so that researchers can observe changes over time in attitudes and perceptions and in the relationships between them.

In his review of survey research methods, Fowler (1993) describes three key methodological components—sampling, interviewing, and question design—that are necessary for designing quality surveys and collecting credible data. First, survey researchers use probability methods to randomly

[2]Two kinds of surveys provide change measures of phenomena over time: repeated cross-sectional surveys, which interview new samples of people (or other sampling unit) at annual or other intervals, and longitudinal or panel surveys, which interview the same people more than one time. Repeated cross sections are useful to construct time series, such as percentages of white and black people who perceive discrimination against blacks; longitudinal surveys are useful for analysis of changes in individual behavior and attitudes and reasons for them.

draw large representative samples from specified populations in order to ensure statistical generalizabilty. Second, they rely on multiple interviewers rather than a single investigator to reduce bias, and they use structured interviews or standardized questionnaires to ensure consistency of data collection and measurement. Third, they design survey questions to gather objective information or to measure subjective phenomena, often by using scale measures.[3] It is essential that the survey questions be reliable and consistent with what the researchers are trying to measure.

When developing survey content, researchers consider the general data collection approach (e.g., personal interviews, computer-assisted telephone interviewing, or mail questionnaires), as well as such factors as question order, content, wording, and type (e.g., open-ended versus fixed responses). As demonstrated by Bobo and Suh (2000; see also Suh, 2000) and Smith (2001), the use of open-ended questions in surveys is particularly valuable for obtaining detailed descriptions of incidents of discrimination. Such questions combine some of the best strengths of surveys (e.g., generalizability and standardized data collection) and in-depth interviews (e.g., narrative richness and detail). Open-ended questions yield specific details about the nature and circumstances of discriminatory behavior that can often validate (or indicate inaccuracies in) direct, self-reported accounts, thereby providing useful information for understanding discrimination.

Key Examples of Surveys

Over the years, researchers have asked thousands of questions and conducted numerous studies on the state and nature of intergroup relations in America (Bobo, 1997; Bobo and Kluegel, 1997; Jackman, 1994; Kinder and Sanders, 1996; Newport et al., 2001; Schuman et al., 1997; Sears and Jessor, 1996; Sears et al., 1997; Smith, 1998, 2000, 2001; Sniderman and Piazza, 1993; Steeh and Krysan, 1996). Common sources of survey data include national polls (Henry J. Kaiser Family Foundation, 1999; Newport, 1999; Newport et al., 2001; Sigelman and Welch, 1991; Smith, 2000; Sniderman and Carmines, 1997; Weitzer and Tuch, 1999); community-based surveys (Bobo and Suh, 2000; Brown, 2001; Gary, 1995; Smith, 1993; Suh, 2000); and specialized studies of employees, employers, professionals, and other target populations (Antecol and Kuhn, 2000; Braddock and McPartland, 1987; Preston, 1998; Supphellen et al., 1997; Yen et al., 1999).

Some widely used surveys of race relations include the Roper polls (Roper Center for Public Opinion Research, 1982), the Multi-City Study of Urban Inequality (Bobo, 1997; Bobo and Kluegel, 1997), the GSS (discussed

[3]Later in this chapter we discuss the use of scales to measure subtle and explicit forms of racism.

further below; Davis et al., 2001), and public opinion polls from the Gallup Organization and Princeton Survey Research Associates (Smith, 2001). The Multi-City Study of Urban Inequality, fielded around 1993 and administered in four urban areas—Atlanta, Boston, Detroit, and Los Angeles—contains explicit measures of racial stereotypes and outgroup perceptions of whites, blacks, Hispanics, and Asians (Bobo and Massagli, 2001; Kluegel and Bobo, 2001; O'Conner et al., 2001). A compilation of holdings of the Roper Center archives in 1982 listed some 4,850 questions dealing with race relations (Roper Center for Public Opinion Research, 1982). Since 1995, the database of national polls maintained by the Roper Center has listed 2,453 questions on the topic of blacks or minorities, at least 359 of which deal with a wide range of items on discrimination, such as the level of discrimination experienced in general public policies dealing with discrimination, personal experiences of discrimination, and causes of discrimination. Surveys of discriminators are generally restricted to employers or other decision makers rather than the general public (e.g., Kirschenman and Neckerman, 1991; Neckerman and Kirschenman, 1991; Supphellen et al., 1997).

Currently, the best example of a large-scale survey of intergroup relations is the GSS, a biennial social indicator survey conducted by the National Opinion Research Center since 1972 (Davis et al., 2001). The GSS receives its core funding from the National Science Foundation and supplemental funding from a wide variety of organizations. The survey uses a national probability sample of 3,000 noninstitutionalized adults to collect information on trends in attitudes and behaviors using a set of complex questionnaires. Each questionnaire includes a standard core of demographic and attitudinal variables, as well as topics of special interest, such as Internet use, multiculturalism, national security, health status, and religion.

Methodological experiments are incorporated into the GSS data collection each year. Such field experiments are also conducted within the other surveys referenced above to assess how the framing of questions affects responses relevant to discrimination. Recent developments in survey research, principally involving computer-assisted telephone interviewing, have made it possible for survey respondents to be assigned easily to random sets or orders of questions (see, e.g., Emerson et al., 2001; Gilens, 1996). Box 8-1 provides an example of such an experiment.

Questions in these surveys are generally used to obtain data on various topics, including stereotypes, social distance, intergroup contact, and discrimination. Among those questions on discrimination in the GSS, for example, are items on personal experiences of discrimination, assessments of discrimination experienced by a particular group, or, less commonly, acts of discrimination committed by the respondent or people he or she knows. However, few of the questions dealing with discrimination provide vali-

BOX 8-1
Experiments in Surveys About Race: An Example

The use of a "split ballot" experiment embedded in a large-scale national survey enables researchers to draw inferences about the general population of responders to surveys with samples that are larger and more diverse than the student population in a typical laboratory experiment. Gilens (1996) used such an experiment within a sample survey to examine whites' racial attitudes and views about welfare using data from the 1991 National Race and Politics Study. The survey respondents were randomly assigned to one of two treatment groups: Half were asked about their beliefs regarding black welfare mothers and half about their beliefs regarding white welfare mothers. Under the assumption that the two groups are interchangeable—guaranteed (on average) by randomization—a difference in the evaluation of welfare provides an estimate of the degree to which race coding may be implicated in white opposition to welfare. The results showed that whites viewed black and white welfare mothers similarly but had more negative views of black welfare mothers when considering policies on welfare. As in laboratory experiments, there may be a difference between responding to questions in a survey and acting in a discriminatory fashion in settings that affect others. Yet as Gilens demonstrates, determining how such attitudinal differences relate to behavioral differences (such as differences in political choices) is important.

dated measures of the occurrence of unequal treatment (see Smith, 2002). The most valuable of these items measures the level of discrimination experienced by individuals or the level of discrimination experienced by particular groups or venues. Questions about the overall level of discrimination without regard to personal experiences, groups, or venues are probably too general to be of much use.

Imperfect Relationship of Attitudes to Behaviors

Smith (2002:14) notes that "many of the questions on intergroup relations in the holdings of the Roper Center or on the GSS provide important information on the state of intergroup relations and help one to understand the context and causes of discrimination, public support for policies to combat discrimination, and related matters." However, relatively few surveys attempt to measure the incidence of discrimination directly at either the individual or the collective level. Smith continues: "Given that only a moderate correlation exists between intergroup beliefs and attitudes (e.g., stereotypes and prejudice) and discriminatory actions (Dovidio, 1993; Dovidio et al., 1996; Fiske, 2000), studying the former is not the same as measuring the latter."

In most studies of attitudes and behaviors, the correlation is typically moderate (Eagly and Chaiken, 1993), and the intergroup domain is no exception. The methodological solutions in the racial discrimination area are the same as in other areas:

- Investigators must create a careful match between the attitude and the behavior. For example, some intergroup experiments (described in Chapter 6) indicate that subtle attitudes best predict subtle (but powerful) behaviors, whereas overt attitudes predict more overt behaviors (Dovidio et al., 2002). As another example, there is evidence that whites' general attitudes toward blacks predict patterns of discrimination, not necessarily any one randomly chosen behavior; by the same token, there is evidence that whites' attitudes toward a specific racial policy or practice (such as busing in a particular school district, housing in a particular neighborhood, or affirmative action in a particular employment or education context) predict their specific behavior in that regard. This elementary point is often missed (Ajzen, 2001).
- The racial attitude–behavior match will depend on moderator variables relating to the type of person involved. Some people act on their attitudes more reliably than do others (Snyder and Swann, 1976).
- The racial attitude–behavior match will also depend on the situation in which the attitude and the discrimination are measured. Social norms strongly affect both what is reported and what is enacted (e.g., Ajzen, 1991), so they can override the ability of attitudes to predict behavior.
- The strength of the attitude itself affects related behavior (Petty and Krosnick, 1995).

Survey Limitations

Smith (2002) describes a number of limitations of surveys. These limitations encompass both methodological factors and reporting biases.

Methodological factors. As noted above, surveys cannot directly measure discrimination; they capture self-reported evidence on perceptions and experiences of discrimination that is not validated. Moreover, such factors as target group, data collection mode, interviewer race, venue, and time of occurrence, as well as question format (e.g., explicit or implicit, phrasing, order), mode (e.g., survey or in-depth interview), and context, can affect the accuracy and completeness of reported perceptions and experiences of discrimination. Discrimination that is subtle or indirect, for instance, may not be readily detected using explicit items. Respondents may also use different meanings for discrimination from one reported account to another (e.g., individual, group, or structural discrimination).

Reporting biases. As the leading historical targets of discrimination in the United States, African Americans may be in the best position to assess the ongoing reality of race in public life. Some scholars, however, suggest that African Americans may have become more sensitive to discrimination because of their socialization since the passage of civil rights legislation, leading them to notice prejudicial actions now more than in the past and producing an upward reporting bias over time (Bobo and Suh, 2000; Brown, 2001; Gary, 1995; Gomez and Trierweiler, 2001; Sigelman and Welch, 1991; Suh, 2000). This reporting bias may also occur if respondents attempt to support a personal conviction that America is a racist society or to explain unfavorable situations and outcomes in their own lives (Harrell, 2000; Lucas, 1994).

Given that blacks are likely to have been well attuned to racial discrimination both before and after the civil rights legislation, however, such upward reporting biases are not likely to be severe. Coleman et al. (2002) actually find that blacks sharply underreport exposure to discrimination when such reports are compared with separate statistical estimates of wage discrimination. In general, researchers have found direct self-reports of discrimination to be accurate and reliable when cross-validated against other data sources (Bobo and Suh, 2000; Essed, 1991; Landrine and Klonoff, 1996). For example, in one recent national survey and one local survey, items on racial discrimination were followed by open-ended questions asking people to describe the mistreatment they supposedly had experienced (Smith, 2000). Across all racial and ethnic groups, some 95 percent of respondents were able to describe specific, appropriate, and credible incidents of discrimination.

Whatever biases may stem from changes in sensitivities among blacks or other racial or ethnic minorities, they are likely to be small in comparison with shifts in white behavior from fairly overt forms to more subtle forms of discrimination. As noted above, because most overtly discriminatory behaviors were declared illegal by the mid-1970s, continued differential treatment on the basis of race has had to become subtler and often may not be apparent even to its targets. Other things being equal, the shift from more overt to more subtle forms of discrimination by whites may actually increase underreporting by blacks because they may be less likely to perceive subtle forms of discrimination. To the extent that real shifts have occurred in white racial attitudes and behaviors, reported declines in discrimination by blacks will be accurate, but part of the decline may reflect the fact that discrimination has been declared illegal by the federal government with the consequence that public expressions of racial prejudice and overt forms of discrimination are widely considered to be socially unacceptable. Annex 8-1 presents survey findings on perceptions of discrimination among both blacks and whites over the past few decades.

Means of Improving Survey Measures

Methodological improvements. Perhaps the most important area for methodological research to improve survey measures of discriminatory experiences and perceptions is research to help understand what aspects of survey and question design may affect levels of reported discrimination, such as the degree to which reports are influenced by race priming.[4] To address this issue, Smith (2002) suggests investigating different approaches to the wording and placement of questions experimentally through random assignment on existing social surveys. If levels of reported discrimination vary by question wording and order, follow-up work will be needed to explain these variations and establish which wordings and placements yield the most accurate results. Another area for research is to use bounded and aided recall techniques in panel surveys to learn how cross-sectional reports may be distorted by errors of memory and telescoping. In addition, cognitive research is needed to understand what respondents specifically include and exclude when they hear such terms as "discrimination" and "unfair treatment."

One fruitful avenue for improvement might be greater use of the factorial vignette method, in which stories are presented to respondents about people being questioned by the police, applying for a job, running for political office, and the like, with the race of the people in the vignette, along with other relevant factors (e.g., age, gender, education, work experience, criminal history), being systematically varied by random assignment. This approach enables researchers to see experimentally how race influences respondents' reported perceptions and evaluations (for a recent example, see Emerson et al., 2001).

Finally, more and better validation studies are needed to discover which approaches yield the most accurate reports of behavior. One possible way of validating self-reports is to collect and compare information from representative samples on reported behaviors and practices that may be consid-

[4]According to Smith (2002), the race-priming hypothesis argues that people will search their memory for negative events and try to assign racial meaning to them either to fulfill the question's request for such incidents or because the cognitive focus on race will color how the respondent reports uncertain or ambiguous events (see Brown 2001; Kinder, 1998).

ered discriminatory (potential discriminators) and on reported experiences with discrimination (targets of discrimination). The former approach may not prove practical, however, if discriminators do not perceive their actions as unfair or biased (Dovidio, 1993; Essed 1991) or if self-recognized acts of discrimination are underreported because of social desirability bias (discussed above). Nonetheless, the approach merits investigation, as incidents of reported discrimination necessarily have both perpetrators and targets. Studies of employers also show that it is possible to collect self-reports of prejudiced attitudes and biased actions against members of disadvantaged racial groups (see Kirschenman and Neckerman, 1991; Supphellen et al., 1997).

Perhaps the most basic form of cross-validation is the use of multiple methods in measuring discrimination. For example, employment discrimination at a company might be studied by examining grievances filed, carrying out surveys of employees and bosses, conducting in-depth interviews, and analyzing employee records.

Other improvements. Smith (2002) offers suggestions for improvements that do not involve additional methodological research. First, surveys should include more target groups. In most surveys, statistically reliable results are available only for whites and blacks, yet Hispanics and Asians are rapidly increasing their shares of the U.S. population, and Arabs and Muslims have recently become prominent as potential targets of prejudice.

Second, surveys should elaborate and extend their measures of discrimination. Some surveys have used only a single measure. Questions need to be refined substantively as well as methodologically to capture subtle and not just explicit discrimination (Dovidio, 1993; Pettigrew and Meertens, 1995; Sears et al., 1997). In this regard, Supphellen et al. (1997) find that projective measures of employment discrimination (e.g., rating the attitudes or opinions of others) are more valid than direct self-reports.

Third, surveys should inquire about discrimination within specific venues and not rely on global questions that leave time and place unspecified. Typical venues included in surveys to date include work, restaurants, stores, interactions with police, schools, housing, public transportation, banks, and government agencies (Brown, 2001; Collins et al., 2000; Gary, 1995; Newport et al., 2001; Smith, 2000). An "other" category might be included to capture incidents occurring outside these venues (Smith, 2000). Specific time periods should also be specified. Doing so would enable estimation not only of the incidence of discrimination but also of its frequency. Lifetime "ever" questions are less than ideal because they demand recall over extended periods of time (which is cognitively difficult and leads to underreporting); they confound cross-sectional monitoring and time-series analysis (because some people report on events that occurred only in the distant past, and others do not); and they underestimate the frequency (as opposed to the incidence) of discrimination (Suh, 2000).

Fourth, not only is it important to measure discrimination, it is also important to assess racial preferment. In 1991, for example, 38 percent of black respondents to the GSS reported that their promotion opportunities were worse because of their race, but 18 percent indicated that their chances were better, yielding a net disadvantage of –20 points. In contrast, white respondents gave responses that yielded a net advantage of +19 points (Smith, 1998).

Governmental Administrative Data

After the passage of the various civil rights acts prohibiting racial discrimination, bureaucratic agencies were established and charged to investigate complaints, monitor compliance, and work to eliminate bias.[5] For example, the National Fair Housing Alliance (NFHA) produces an annual report on the state of housing discrimination in the United States that is based on information from local fair housing groups, the U.S. Department of Housing and Urban Development, the U.S. Department of Justice, and numerous state and local government agencies. Discrimination is tracked in four different market sectors: rental markets, mortgage lending, home sales, and homeowner insurance. Complaints filed with NFHA and other agencies provide one source of data on the underlying trends in the frequency and incidence of housing discrimination. The NFHA data compiled for 2001 show that race continues to be the most commonly reported basis for discrimination, accounting for 32 percent of the 23,557 cases filed (National Fair Housing Alliance, 2002). Other potential sources of administrative data include antidiscrimination suits filed in state and federal courts (Garrett, 2001; National Research Council, 1989; Romero, 2000; Shivley, 2001) and registries of hate crimes maintained by state and local human rights commissions and race relations boards (Evans, 2001; Strom, 2001).

As noted by Smith (2002), administrative data have several advantages: They represent socially significant events, they are publicly accessible, and they are generally inexpensive to use because the costs of collection are borne by the enforcing agency. Using industry data from the Bureau of Labor Statistics, for example, Petersen and Morgan (1995) were easily able to document the skewed placement of men and women within occupational

[5]These agencies include local fair housing and employment commissions and federal agencies, such as the U.S. Equal Employment Opportunity Commission, the U.S. Commission on Civil Rights, the Office of Fair Housing Enforcement at the U.S. Department of Housing and Urban Development, and the Office of the Assistant Attorney General for Civil Rights (National Research Council, 1989; U.S. Equal Employment Opportunity Commission, 2002).

categories within firms. The data showed that women were far more likely to be in poorly paid jobs. This finding suggests that if one were to look for possible gender discrimination, it would be important to look not just at wage differences within jobs but also at the mechanisms that allocate men and women to occupations within the firm.

As Smith points out, however, relying on government reports also has serious limitations. First, the data are collected to meet legal requirements, not to facilitate social science research. In other words, coverage is defined by law, and the information is collected largely for administrative rather than scientific purposes. Second, willingness to report discriminatory treatment may depend on the ease of reporting and the vigor with which an agency deals with complaints (Lucas, 1994). As these conditions vary over time, trends may be biased (Shivley, 2001). Finally, not all discriminatory practices are illegal (and thus covered by government enforcement mandates), and many acts of discrimination that are now illegal were not so several decades ago (Romero, 2000). The Anti-Defamation League (2001), for example, includes acts of anti-Semitic speech in its annual report on anti-Semitism, even though such speech is legal. Because of problems with the quality and completeness of administrative data, the usefulness of these data sets for research purposes is limited.[6]

Nongovernmental Data

In addition to government agencies, many private organizations maintain a record of discriminatory complaints. However, complaints in such cases are made to internal grievance boards governed by formal procedures within an organization; an example is complaints filed with an employer or labor union. Internal data from for-profit organizations have many of the same limitations as those from governmental agencies. Companies establish procedures to serve legal and business purposes, not to further a scientific research agenda. In addition, company reports are typically not made public, and their nature and specifics vary considerably across organizations.

Racial grievances may also be lodged with independent nonprofit organizations whose mission is to promote racial or ethnic equality, such as the Anti-Defamation League or the Mexican-American Legal Defense and Educational Fund. Reports to nongovernmental third parties have the advantage that they are publicly accessible, typically cover many different types of discrimination in various venues, and draw from across the nation. However, the resulting data are often compiled by highly self-interested organizations that are not scientifically motivated. Moreover, as with government

[6]See Ross and Yinger (2002) for a detailed discussion of issues that arise for scholars and enforcement officials when collecting and analyzing data to study discrimination.

and other organizational reports, these sources rely on targets of discrimination knowing where and how to report mistreatment and being motivated and able to do so.

In-Depth Interviews

In conducting in-depth or qualitative interviews, a researcher engages one or more subjects in an extensive, semistructured conversation, which is often audio recorded (Bonilla-Silva and Forman, 2000; Essed, 1997; Feagin, 1991; Feagin and Sikes, 1994; St. Jean and Feagin, 1998, 1999; Zweigenhaft and Domhoff, 1991). The advantages of such interviews are that they ask individuals about their actual experiences of discrimination and often elicit information that is richly detailed. Essed (1991), for example, claims that using unstructured interviews rather than highly structured ones allowed her to draw out more detailed accounts of people's experiences. She also notes that participants were able to express intuitive feelings regarding their experiences with prejudice and discrimination that might otherwise be difficult to articulate and to report events in ways that others might consider oversensitive.

In-depth interviews are generally based on small, usually unrepresentative samples that are often biased because participants are of higher status, more articulate, and more politically aware than most of the subject population (see Essed, 1991; Feagin, 1991; Zweigenhaft and Domhoff, 1991). The most frequently used method of selection is the chain referral method, also known as snowball sampling, in which respondents provide leads to other potential interviewees within their social networks. Because social networks tend toward homogeneity, snowball sampling can yield biased samples and findings that are difficult to generalize (Goodman, 1961). Moreover, the accuracy of self-reports may be affected by ways in which respondents react to interviewer probes for reports of discrimination or other behaviors, or if respondents provide vague and poorly detailed accounts (Smith, 2002).

SCALE MEASURES USED IN SURVEYS

As discussed above, although survey-based self-reports have been found to be reliable, accurate, and useful ways of measuring experiences of discrimination, the shift from overt to subtle forms of discrimination has made it more difficult to assess the occurrence of discrimination or to capture people's beliefs using survey questions. Many observers have noted an ongoing conflict between principle and practice regarding racial prejudice in the American psyche and have theorized the contradiction under a variety of conceptual rubrics (Fiske, 1998): modern racism (McConahay, 1986), symbolic racism (Kinder and Sears, 1981), ambivalent racism (Katz et al., 1986), aversive racism (Dovidio and Gaertner, 1986), laissez-faire racism (Bobo et al., 1997; Bobo and Smith, 1998), and subtle racism (Pettigrew and Meertens, 1995) (see also Annex 8-1). Although individual survey questions may not capture such complex attitudes and ambivalence about race, scale measures, which combine information from multiple questions, can capture the sometimes contradictory mechanisms that lead to more subtle types of discrimination. We turn to these measures next.

Measures of Modern Racism

The psychological source of a conflict between racial equality in principle and practice results from people's assimilation of egalitarian laws and norms, along with their internalization of continued cultural messages that blacks are inferior to whites. For example, many whites, regardless of explicit individual prejudices, make rapid associations between whites and positive ascriptions and between blacks and less positive ascriptions (e.g., Devine, 1989; Dovidio et al., 1986; Gaertner and McLaughlin, 1983; see Chapter 6). In part, these associations may reflect media coverage depicting blacks less favorably than whites in news reports and in entertainment. Many Americans resolve the conflict between these culturally influenced associations and the nation's egalitarian values by maintaining those values simultaneously with subtle, automatic, or indirect forms of prejudice (see Chapters 4 and 6).

In survey settings, measures of modern racism examine reactions to black Americans as perceived threats to whites' traditional values and economic status (Duckitt, 2001). The single most commonly used measure since the mid-1970s has been McConahay's (1986) Modern Racism Scale—an outgrowth of work on symbolic politics (Kinder and Sanders, 1996; Kinder and Sears, 1981). Generally split into three parts, this scale typically encompasses modern (subtle) racism, old-fashioned (overt) racism, and filler items on irrelevant current events to disguise the measure. Items on modern racism include believing, for example, that the government and the media give blacks more respect than they deserve, that discrimination is no longer a problem, that blacks have gotten more economically than they deserve, and that blacks are too demanding. In contrast, items on old-fashioned racism ask about open opposition to fair housing laws, integration, intermarriage, and having a black neighbor.

Scores on the Modern Racism Scale predict a variety of variables related to discrimination to varying degrees (Kinder and Sanders, 1996). Examples include conscious endorsement of stereotypes about black Americans (Devine, 1989), anti-black feelings, simulated hiring decisions, voting for white over black candidates (even after controlling for ideology), and opposing busing to achieve school desegregation (even after controlling for related variables). A variant of the scale (Pettigrew and Meertens, 1995) distinguishes subtle and explicit prejudice against immigrant minorities in Europe. Items on this scale include blaming minorities for being too pushy and intrusive but simultaneously not trying hard enough, as well as withholding sympathy for their disadvantaged situation.

Although widely used, the Modern Racism Scale has provoked two kinds of scholarly debate. First, while few dispute the empirical correlates of the scale, their meaning is open to interpretation, either as a perceived threat to the situation of one's own group or as principled conservatism. For example, some argue that opposition to racial integration, school desegregation, affirmative action, and welfare programs reflects primarily opposition to government intervention (Sniderman, 1985); however, the role of such ideology appears to be empirically small (Kinder, 1998). Some argue that racial attitudes reflect narrow economic self-interest (Bobo and Kluegel, 1997); that is, people vote their own personal wallet. Yet perceptions of the economic standing of one's racial group as a whole appear to be at least as strong a predictor of racial attitudes and voting behavior as perceptions of one's own economic standing (Kinder and Kiewiet, 1981; Sears and Funk, 1991). This suggests that modern racism is based on the interests of one's group, not one's own self-interest. Moreover, consistent with the idea of modern racism, the importance that white Americans attach to equality of opportunity, relative to their other core values, correlates with their more specific racial attitudes and policy preferences (Kinder and Sanders, 1996). In other words, modern racism represents a syndrome of perceived own group interest, basic values, and specific attitudes toward policies—all of which disadvantage minorities.

Some common ground for this first type of debate—over the meaning of modern racism as group threat versus principled conservatism—appears in findings that modern, subtle forms of prejudice correlate with (a) subjectively perceived threats to group status (Bobo, 1983), (b) positive feelings toward one's own racial group, and (c) negative feelings toward racial outgroups (Wood, 1994). Group prejudice may not be the only factor predicting racial attitudes and policy preferences, but as one commentator notes, "it is always present, and of all the ingredients that go into opinion, it is often the most powerful" (Kinder, 1998:805).

A second form of scholarly debate has emerged more recently. The original items on the Modern Racism Scale were considered relatively nonreactive; that is, respondents did not necessarily interpret them as measures of discrimination and so apparently did not monitor their responses for social desirability. Over the past decade, however, growing evidence suggests that the items have become reactive (Fazio et al., 1995). Modern racism scores now correlate with more explicit measures of deliberative discrimination, rather than with automatic or implicit measures (e.g., Dovidio et al., 1997), although correlations with more implicit, spontaneous measures of discrimination still occur (Wittenbrink et al., 1997, 2001).

More generally, various dimensions of prejudice matter to understanding both prejudiced attitudes and discriminatory behavior. One dimension contrasts spontaneous, implicit, automatic, and subtle versus deliberative, explicit, and controlled reactions (see Chapter 6, Box 6-3); another important dimension contrasts emotional, evaluative, and affective processes versus belief, conceptual, and cognitive processes (see Chapter 6 and Talaska et al., 2003); and a third dimension contrasts ingroup preference versus outgroup derogation (see Chapter 4).

One key component of measuring subtle racism is that it depends heavily on context. As an example, to the extent that white attitudes are ambivalent—encompassing both sympathy and rejection—people may act on different aspects of their attitudes under different circumstances. Racial ambivalence (Katz and Hass, 1988; Katz et al., 1986) suggests the co-occurrence of blaming anti-black feelings (the perceived irresponsibility of black families, leaders, and values underlies the continuing disadvantage of black Americans) with paternalistic pro-black feelings (emphasizing obstacles, discrimination, and unequal opportunities). The work by Katz and colleagues demonstrated empirically that anti-black attitudes correlated with white perceptions that blacks violate values related to the Protestant work ethic and that pro-black attitudes correlated with humanitarian and egalitarian values. Whites can simultaneously possess both sets of attitudes. The implication for discriminatory behavior is that reactions toward a single black individual can be affected by a small push in either a positive or negative direction (e.g., slightly superior or slightly inferior credentials for a job applicant). Racially ambivalent whites then overreact, making excessively positive or excessively negative decisions as compared with their decisions about a comparable white individual. The Modern Racism Scale elicits a similarly exaggerated response, either overly positive or overly negative, in simulated hiring decisions.

The common thread of all the work on modern racism, symbolic racism, subtle prejudice, and ambivalent racism is that appearing racist has become aversive to many white Americans (Dovidio and Gaertner, 1986), creating more complex racial attitudes than in the past. Whatever their source, subtle, modern forms of racism predict avoidance of interactions with racial outgroups, and such passive rejection can result in discrimination (see Chapter 3).

Survey research would be improved by a more explicit model of what forces determine expressed attitudes: expectations, experiences, social pressure, and so on. The analysis of relationships between past experiences, expressed attitudes, and future behavior seems remarkably undeveloped.

Embedding attitude reports in a lifetime of behaviors would permit this area of research to develop in new ways. Furthermore, research on modern racism can inform statistical and experimental research by examining relevant individual characteristics (e.g., personal values) and features of encounters (e.g., interdependence, accountability). Some social science models, such as the theories of reasoned action and planned behavior, provide a framework for incorporating a variety of factors to predict behavior and evaluating the respective weights and values of own beliefs, perceived social norms, and perceived behavioral control, as well as own prior behavior (Ajzen, 2001).

Measures of Explicit Racism

Not all white Americans eschew overt racial bias. By some estimates, about 10 percent of white Americans openly embrace racial discrimination; if accurate, this figure would mean there is at least one prejudiced white person for every black person in the United States (Fischer et al., 1996). Hence, even this apparently low incidence of prejudice can be viewed as high. Moreover, measures of more explicit racially biased attitudes correlate with more explicit and potentially violent forms of discrimination. Not everyone holding these attitudes is violent, of course, but perpetrators of violence do hold these attitudes. Thus, explicit racism is a necessary but not sufficient predictor of the most serious forms of discrimination.

Three contemporary scales predict endorsement of overtly prejudiced attitudes. One such scale measures blatant or explicit prejudice (Pettigrew and Meertens, 1995), defined as resentment of racial and ethnic minority groups (their allegedly stealing ingroup jobs but also using welfare), as well as rejection of ties to minorities (having a mixed grandchild or an outgroup boss). High scores on this scale predict generalized ethnocentrism and overall rejection of outgroups not one's own, as well as approval of racist political movements and hate crimes (Green et al., 1999). Explicit prejudice stems from perceived threat to the economic status of one's own group.

A related scale measures social dominance orientation (Sidanius and Pratto, 1999); it also focuses on economic and status competition between societal groups. High scores on this scale indicate agreement by respondents that some groups are just more worthy than others, that group hierarchy is inevitable and good, and that dominance is necessary. These attitudes correlate with believing that the world is competitive and that force is sometimes necessary to keep inferior groups in their place. They also correlate with explicit racial prejudice.

In addition to perceived economic threat, perceived threat to values predicts discriminatory attitudes. A scale on right-wing authoritarianism (Altemeyer, 1988, 1996) measures belief in old-fashioned ways, censorship, leadership based on superior power, and rejection of troublemakers and deviants. High scores on this scale correlate with social conservatism and predict approval of aggression against nonconformers. They also correlate with limited intergroup contact and limited education.

SUMMARY AND RECOMMENDATIONS

Following the enactment of civil rights legislation, overt racial discrimination became illegal and socially unacceptable, and measuring discrimination became increasingly difficult as a result. The more subtle forms of discrimination evident today complicate the way we assess the causes and consequences of discrimination. Surveys provide valuable evidence for understanding the extent of discrimination; however, they cannot directly measure its occurrence. Most survey items are intended to measure self-reported attitudes, perceptions, or experiences of discrimination, and these items can be unreliable for at least two reasons. First, if a discriminatory occurrence is ambiguous, a black or other minority respondent may under- or overreport its incidence. Some subtle forms of discrimination, for instance, may not be easily detected. Second, white respondents are often not willing to admit to practicing or supporting discrimination, which may lead to less-accurate reporting of their true attitudes or beliefs.

On the other hand, although survey measures cannot capture discrimination directly, results to date suggest that valid and reliable data on racially discriminatory attitudes and experiences can be gathered on social surveys. Conducting repeated cross-sectional surveys is very useful to provide time series; the GSS is the best example of a large-scale survey that collects data on changes over time in racial attitudes and experiences with discrimination through yearly interviews with samples of the population. Conducting longitudinal surveys to analyze the incidence, causes, and consequences of changes in attitudes about race and experiences of racial discrimination at the individual level is also very valuable, although none of the major longitudinal surveys to date has included attitudinal or perceptual variables (see further discussion in Chapter 11). More generally, research is needed to evaluate and improve the reliability and consistency of survey reports of discriminatory attitudes and behavior. In particular, as expressions of prejudice and discriminatory behavior change and become more subtle, survey questions on racial attitudes and experiences of discrimination may be necessary. Scale measures can be very useful to capture complex racial attitudes that can lead to more subtle types of discrimination. Open-ended questions on surveys can provide some of the advantages of in-depth interviews in regard to the detail of information obtained combined with the advantages of surveys of large, representative samples.

Reports of discrimination in administrative records systems, such as those of government civil rights enforcement agencies, private organizations, and nonprofit groups, can also provide useful information for analysis that is available at low additional cost to the researcher. However, some administrative records may be difficult to obtain by researchers, the reporting of events may be biased in several ways (e.g., by changes in the vigor with which an agency pursues enforcement of antidiscrimination laws and policies), and data on covariates may be very limited, thereby restricting the use of the records for research as well as administrative purposes.

Recommendation 8.1. To understand changes in racial attitudes and reported perceptions of discrimination over time, public and private funding agencies should continue to support the collection of rich survey data:

- **The General Social Survey, which since 1972 has been the leading source of repeated cross-sectional data on trends in racial attitudes and perceptions of racial discrimination, merits continued support for measurement of important dimensions of discrimination over time and among population groups.**

- **Major longitudinal surveys, such as the Panel Study of Income Dynamics, the National Longitudinal Survey of Youth, and others, merit support as data sources for studies of cumulative disadvantage across time, domains, generations, and population groups. To further enhance their usefulness, questions on perceived experiences of racial discrimination and racial attitudes should be added to these surveys.**

- **Data collection sponsors should support research on question wording and survey design that can lead to improvements in survey-based measures relating to perceived experiences of racial discrimination.**

Recommendation 8.2. Agencies that collect administrative record reports of racial discrimination should seek ways to allow researchers to use these data for analyzing discrimination where appropriate. They should also identify ways to improve the completeness, reliability, and usefulness of reports of particular types of discriminatory events for both administrative and research purposes.

ANNEX 8-1: BLACK AND WHITE AMERICANS' PERCEPTIONS OF DISCRIMINATION AND AMBIVALENT ATTITUDES ABOUT RACE

Black Perceptions of Discrimination

Prior to the passage of civil rights legislation, blacks perceived the United States to be a highly prejudiced and discriminatory society. In 1963, more than three-quarters of a nationally representative sample of black Americans perceived significant racial discrimination in U.S. job markets (Schuman et al., 1997). Asked whether they had as good a chance as whites to get jobs for which they were qualified, a resounding 77 percent said "no." In the following years, the percentage of black respondents perceiving employment discrimination fell, reaching 64 percent in 1978 and 55 percent in 1989. As late as 1997, more than half (53 percent) of all African Americans said that blacks still did not have as good a chance as whites to get jobs for which they were equally qualified. Moreover, in 1996, 63 percent of African Americans nationwide continued to view discrimination as a primary cause of disadvantage among blacks (Schuman et al., 1997).

Between 1997 and 2001, the Gallup Organization and Princeton Survey Research Associates public opinion polls asked nationally representative samples of African Americans to report any discrimination or unfair treatment they had experienced within the past 30 days (Smith, 2001). On average, 26 percent of respondents said they had experienced discriminatory treatment while shopping, 16 percent at the workplace, and 16 percent while on public transportation. A national survey of African Americans sponsored by the *Washington Post* during 2000 found that 30 percent had at least sometimes been "called names or insulted" and 17 percent had been "physically threatened or attacked" because of their race. Rates for nonblacks were substantially lower, with only 18 percent of the general population reporting a racial or ethnic insult and 11 percent a physical threat or attack. Clearly, African Americans, more than other Americans, still perceive significant discrimination in public life and view it as a significant barrier to their social and economic advancement.

In the Multi-City Study of Urban Inequality, each respondent was asked: "In general, how much discrimination is there that hurts the chances of blacks to get good-paying jobs?" In Atlanta, 60 percent of black respondents answered "a lot," compared with 57 percent in Boston, 62 percent in Detroit, and 69 percent in Los Angeles. When respondents who answered "some" to the same question were added, the percentage of African Americans who perceived racial discrimination in job markets rose to well over 90 percent in each metropolitan area (Kluegel and Bobo, 2001). Of course, the implicit definition of discrimination used by respondents may not be the same as the legal definition currently recognized by U.S. courts.

White Perceptions of Discrimination

Not long ago, white respondents were willing to admit their support for racial discrimination to survey researchers. In the early 1940s, for example, 68 percent of whites nationwide said they thought blacks and whites should attend separate schools, 55 percent said whites should have priority over blacks in hiring, and 54 percent agreed that separate sections should be reserved for blacks and whites on buses and streetcars. By the 1960s, such segregationist attitudes had moderated considerably, with 30 percent of whites still favoring racially separate schools and 11 percent approving of white preferences in hiring. In 1970, 12 percent of whites admitted to favoring racial segregation in public transportation (Schuman et al., 1997).

In the years since the civil rights era ended, these percentages have fallen even further; fewer and fewer whites are willing to express open support for racial discrimination. By 1995, only 4 percent reported they believed that blacks and whites should attend separate schools, just 13 percent said there should be laws against black–white intermarriage (as late as 1963 the percentage was still 62 percent), and over 90 percent of whites endorsed the principle that blacks have a right to live wherever they can afford to live (Schuman et al., 1997). Although white support for racist principles had fallen to low levels, it had not disappeared entirely. As late as 1993, 15 percent of whites agreed that blacks should respect the rights of whites to exclude blacks from their neighborhoods if they so desire (Schuman et al., 1997).

Moreover, while fewer whites openly support principles of racial discrimination, many remain ambivalent in their attitudes about race. For instance, surveys show that only 13 percent of whites support a ban on black–white intermarriage, whereas 33 percent still disapprove of the practice personally. A mere 2 percent object to sending their children to a school where "a few" students are black, while 19 percent object to sending their children to one where half are black. Likewise, just 2 percent of whites said they would move out of their home if black neighbors moved in next door, but 25 percent said they would leave if blacks entered their neighborhood "in great numbers." As of 1995, nearly a quarter of whites (23 percent) said they would object to having a black dinner guest (Schuman et al., 1997).

Ambivalent Attitudes About Race

At the end of the twentieth century, open support for principles of racial discrimination had fallen to very low levels among whites, with only 10 to 15 percent endorsing discriminatory actions or policies. Although whites may have come to support a nondiscriminatory society *in principle*, however, they remain substantially uneasy about its implications *in practice*.

White respondents, for example, continue to express reservations about racial mixing in social institutions where such mixing formerly did not occur, and racial segregation generally remains high in schools, churches, neighborhoods, and marriages (see Emerson and Smith, 2000; Farley, 1996; Massey and Denton, 1993; Orfield and Eaton, 1996). For instance, although most whites now agree that blacks should be able to live wherever they choose, they still want blacks to choose to live somewhere else.

Racial attitudes of individual Americans may not correspond with their views about or support for public policies or practices. Loury (2002) draws a useful distinction between the "social meanings" associated with racial classification and the "racial attitudes" held by individuals. Social meanings refer to the unexamined beliefs that influence how citizens understand and interpret the images they glean from the larger social world. For example, the meaning of a policy regarding job preference may be sensitive to the race of those affected: Veterans have been seen as acceptable beneficiaries, whereas the application of such a policy for blacks has been thought to violate meritocratic principles (Skrentny, 1996). Likewise, views about welfare policies may depend on the race of local recipients, so that more blacks being on the local rolls is associated with greater hostility toward recipients (Luttmer, 2001).

Sniderman and Piazza (1993) illustrate the difference between racial attitudes and racial meanings in their "mere mention" experiment. As noted by Loury (2002), these survey researchers found that "the mere mention" of affirmative action, in the context of soliciting from white respondents their views about racial stereotypes, made those whites more likely to agree with negative racial generalizations (e.g., most blacks are lazy). Compared with two groups of similar whites, the ones to whom affirmative action was "merely mentioned" showed a significantly higher tendency to affirm negative stereotypes about blacks than did those to whom affirmative action was not mentioned at all. The researchers concluded that the respondents' expression of these anti-black sentiments had been "caused" by their dislike of affirmative action, and not the other way around. That is, whites were expressing primarily their ideological views about policy, which, when affirmative action was mentioned, then spilled over to affect their views about race.[7]

[7]Loury (2002) argues that this may be an incomplete interpretation. The whites in Sniderman's experiments may, as he argues, have been driven mainly by ideology and not by racial animus. However, it remains the case that the ideological meanings of a contested racial policy such as affirmative action are determined within a racial context. A similar policy with a different set of beneficiaries might not have the same ideological resonance. Public responses to a social malady, such as drug involvement, may depend on the race of those suffering the problem; thus young urban drug dealers elicit a punitive response, while young suburban drug buyers call forth a therapeutic one (Tonry, 1995).

Regardless of how it is conceptualized, the discrepancy between the acceptance of nondiscrimination in principle and the discomfort about its implications in practice can be traced, at least in part, back to the persistence of anti-black stereotypes. In the mid-1980s, 61 percent of whites nationwide said that blacks on welfare could get a job "if they really tried," 42 percent said that black neighborhoods are more rundown because blacks "don't take care of their property," and 43 percent endorsed the view that blacks would be as well off as whites "if they would just try harder" (Sniderman and Piazza, 1993; Loury, 2002). As of 1991, moreover, 34 percent of whites described blacks as "lazy" and 21 percent labeled them as "irresponsible" (Sniderman and Piazza, 1993). Likewise, according to the 1993 Multi-City Study of Urban Inequality, whites perceived blacks to be significantly less intelligent, less rich, less self-supporting, and less easy to get along with than Asians, Hispanics, or whites (Bobo and Massagli, 2001), and whites perceived African Americans as the least desirable potential neighbors (Charles, 2001).

9

An Illustration of Methodological Complexity: Racial Profiling

We end Part II with a specific example of an area for which research on the role of racial discrimination is important but difficult to carry out. The example we use is racial profiling. Given the challenges to measurement, we do not endeavor to prescribe state-of-the art methods for determining when racial profiling exists or its effects. Rather, our discussion of specific issues regarding methods and data is intended to remind researchers, policy makers, and the public of the difficulties of causal inference with regard to profiling, which may also be relevant for other areas in which racial discrimination may occur.

We begin with definitions of profiling and racial profiling. Profiling is a statistically discriminatory screening process in which some individuals in a population (e.g., automobile drivers, income tax filers, people going through customs, people boarding an airplane) are selected on the basis of one or more observable characteristics and then investigated to determine whether they have committed or intend to commit a criminal act (e.g., sell or smuggle drugs, cheat on taxes, blow up an airplane) or other act of interest. The particular characteristics used in profiling are chosen with the goal of selecting people who are most likely to warrant further investigation and typically depend on the setting. For example, people who purchase one-way airline tickets using cash on the day of their flight may be selected for further scrutiny by airport personnel based on an assumption or empirical evidence that they are more likely than others to pose a risk of premeditated violence to passengers.

We reserve the term "profiling" for screening situations in which there is reason to believe that criminal behavior could be committed, but there is no specific knowledge of a particular suspect or criminal scheme.[1] We thereby distinguish profiling from situations in which a specific description of a suspect is issued on the basis of presumably reliable information.

Racial (or ethnic) profiling is a statistically discriminatory screening process in which race (or ethnicity) is used as one, or the only, observable characteristic in the profile. The problem of racial profiling in law enforcement has attracted a great deal of public attention in recent years. Such profiling is conceptually no different from the kinds of discrimination previously discussed in this report (see Chapter 4); it is simply one instance of the more general phenomenon we have termed statistical discrimination. Racial profiling in the criminal justice arena entails the use by law enforcement personnel of statistical generalizations about a group of people based on their race. To the extent that these generalizations reflect overt racial prejudice or issue from subtle, race-influenced cognitive biases, profiling is indistinguishable from the explicit prejudice we have already discussed. Even when race-based generalizations are consistent with one reading of the evidence (as when, in a certain locality, police officers give heightened scrutiny to blacks because they know that in that locality and on average blacks are more likely than whites to be involved in certain kinds of crime), it remains the case that profiling is a type of statistical discrimination. Thus, our earlier discussion of statistical discrimination based on race also applies to racial profiling.

Earlier we noted that it is unlawful to judge an individual job applicant on the basis of the average characteristics of the applicant's racial group, regardless of whether the employer's assessment of the racial average is accurate (see Chapter 4). Similarly, most observers believe it is wrong for domestic law enforcement personnel to base their routine treatment of individuals on the average behaviors of racial groups. Thus, the results of a Gallup poll in 1999 showed that 81 percent of Americans did not approve of racial profiling, defined as the practice by police officers of stopping drivers from certain racial or ethnic backgrounds because officers believe these groups are more likely to commit certain crimes (Gallup Poll, 1999). There have also been many policy statements by police officials and legislative bodies declaring the unacceptability of racial profiling in police work.[2] Recently, the Bush administration issued policy guidance on racial or ethnic

---

[1]For concreteness, we refer to profiling with reference to a criminal act, but the term applies to screening to detect any activity of interest.

[2]See, for example, National Conference of State Legislatures (2002); Minnesota's statute on racial profiling (http://www.aele.org/minnprofile.html [accessed January 29, 2004]); and Tulsa Police Department Policy 31-316B (http://www.tulsapolice.org/racial_profiling_policy.html [accessed June 9, 2003]).

profiling forbidding its use in federal domestic law enforcement: "'Routine patrol duties must be carried out without consideration of race,' the Justice Department policy states. 'Stereotyping certain races as having a greater propensity to commit crimes is absolutely prohibited'" (Allen, 2003:A14). The only instance in which domestic law enforcement officers may use race is when it is part of a specific description obtained from a witness or informant about a specific crime.

Even when statistical profiling is not explicitly racial, to the extent that it relies on characteristics that are distributed differently for different racial groups, the result may be to produce a racially disparate impact. For example, if the police tend to stop cars with broken tail lights more frequently and if disadvantaged racial groups are more likely to drive older cars, then the profile—stop cars with broken fixtures—will result in a higher stop rate for these groups. Recall that in the employment context the use of screening criteria having a disparate impact on a protected racial group is legitimate only if the employer can demonstrate an objective and suitably compelling connection between the screening criteria and the employer's economic bottom line. So, too, in the context of law enforcement, nonracial profiling that relies on traits distributed differently among racial groups and that results in a racially disparate impact must be justified by demonstrating an objective association between those traits (e.g., broken tail lights) and the outcome of interest (criminality). This would be the case, for example, if it could be shown that drug couriers typically drive older cars (e.g., because they are poorer or because their cars would be confiscated if they were caught carrying drugs).

In this chapter we discuss racial profiling primarily in the context of measurement—that is, how it may be possible to determine when racial profiling is (or is not) occurring in law enforcement. Allegations of discriminatory racial profiling—mainly by police making traffic stops—have increased in frequency in the past few decades.[3] Yet methods and data with which to establish that disadvantaged racial groups are being stopped at higher rates than others and that racial profiling explains some or all of the differences in selection rates are not well developed. The measurement and modeling issues are similar to those discussed in Chapter 7 on using statistical models with observational data to measure discrimination by inference, but some special issues in the profiling situation warrant attention.

We also briefly discuss racial or ethnic profiling as a policy option in the context of the increased threats to public security from terrorist attacks. The panel began its deliberations scarcely one month after the attacks of

[3]Issues of racial profiling in other settings, such as inspection by customs officials for carrying of drugs or other contraband, have also attracted political and legal attention (see Harris, 1997, 1999a; *Washington Post*, 2002; Webb, 1999; White, 2000).

September 11, 2001, so we could not help but be aware of how public discussion and perceptions regarding profiling had changed. Hence, we deemed it of value to discuss the issues involved in the possible use of racial or ethnic profiling (or profiling using characteristics that correlate highly with race or ethnicity) as a tool with the potential to help prevent future terrorist attacks. Some issues are technical, involving how or whether one could determine the potential effectiveness of race, ethnicity, and other characteristics as profiling factors. Other, even more important, issues involve the heavy societal costs of using race or ethnicity (or variables highly correlated with them) in profiles.

Of course, time has passed since we began our deliberations, and public officials, as well as the nation as a whole, have continued to discuss and debate the pros and cons of profiling in the terrorism context. We have not been connected to those debates and do not comment on specific rulings or positions that have been proposed or adopted in the interim (e.g., the Bush administration policy guidance that permits ethnic profiling in narrow circumstances involving international terrorism). Our deliberations were concerned with the general issue of racial or ethnic profiling—how to determine when and whether it occurs in situations when one would want to prevent it and what considerations might need to be taken into account if one wanted to implement it even though it is, by our definition, discriminatory. Although we have not deliberated about and have no comment on specific profiling proposals, we hope the general points we raise will serve to aid public evaluation of the issues.

MEASUREMENT ISSUES

Two main measurement issues arise in attempting to establish the existence of racial profiling in a law enforcement situation. The first is how to determine that racial or ethnic groups are being subjected to enforcement actions (e.g., traffic stops, searches, citations, arrests) at disparate rates. The second is how to determine that racial profiling is a causal factor in disparate selection rates. The discussion here addresses primarily the first issue; the second presents modeling challenges similar to those discussed in Chapter 7 on measuring racial discrimination in labor markets and other settings.

Establishing Disparate Outcomes in Profiling Situations

Data Sources on Racial Profiling

Much of the available data on racial profiling come from anecdotal experiences of nonwhites. In a typical case, a nonwhite person may be pulled over for a minor traffic violation (e.g., speeding 5 miles over the limit) and searched on suspicion of carrying contraband. Similarly, a nonwhite person may be stopped and questioned for being in a predominantly white neighborhood (see Harris, 1999b). Although these incidents are clearly discriminatory, such complaints do not prove that police officers and security personnel engage in racial profiling generally, or even that members of minority groups are necessarily detained more often than others. However, the substantial number of complaints occurring in certain types of situations (e.g., traffic stops) indicates how widely racial profiling is believed to be—and could in fact be—used.

A second source of data on racial profiling is official records, such as state and local police data on traffic and pedestrian stops, searches, warnings, citations, and arrests. Many states, including Maryland, New Jersey, and North Carolina, have enacted legislation for the collection of detailed data on stops and have mandated studies of racial profiling.[4] Despite these efforts, however, relatively few data sets are complete, accurate, and available for analysis (Glaser, 2003; Harris, 1999b). For example, police records on stops may not include the race of those individuals stopped but not cited or arrested by police, and there may be little consistency in reporting race for a variety of reasons.

An important use of detailed police data on traffic stops is to provide early warning of individuals who engage in inappropriate racial profiling. This use can be fraught with danger, however, if the data do not reliably indicate such behavior. One obvious concern is that officers may manipulate their reports if they perceive they are in danger of disciplinary action. Or if they stop members of disadvantaged groups on the basis of race, they may make unnecessary stops of advantaged groups to balance their "portfolio." These corrective actions may keep the record clean but are inefficient as well as discriminatory against the members of such groups. On the other hand, police officers who make appropriate stops may in some cases face unwarranted charges of racial profiling if their stop rates by race are compared with population (baseline) rates that are poorly measured (see below). To the extent that official data are biased in any of these ways because of their use for individual disciplinary actions, the data will also be biased for research purposes.

Yet another source of data on profiling is direct observation of selection decisions. It may be possible for researchers to collaborate with officers in the field to elicit information on what factors they take into account as

[4]For state legislation mandating data collection and other efforts, see Institute on Race and Poverty (2001), National Conference of State Legislatures (2002), and Police Foundation (2001).

they make stop decisions. One could then examine the consistency in those factors across different officers and between the decisions made when accompanied by a researcher and those made when officers are on their own. Such studies must be carried out carefully to avoid biasing the results by virtue of the direct involvement of the researcher.

Methods for Estimating Disparate Selection Rates

Regardless of how complete or accurate the data collected on such law enforcement actions as traffic stops or customs searches may be, those data are likely not to be sufficient in and of themselves to establish the existence of racially disparate outcomes. For example, a finding that more blacks are stopped than whites at a certain intersection may reflect the fact that more black drivers pass through that intersection (because of residential or employment isolation) than do white drivers.

Indeed, the most common problem cited across studies of police profiling (e.g., Engel et al., 2002; Fagan, 2002; Lamberth, 1994, 1996; Ramirez et al., 2000; Zingraff et al., 2000) is identifying the appropriate population to classify by race for comparison with the racial classification of those stopped by the police—the so-called denominator or base rate problem. For example, if one has police data on the percentage of nonwhites stopped at an intersection among all people stopped, is an appropriate comparison measure the percentage of nonwhites in the population living around that intersection, the percentage of nonwhites observed to drive by that intersection on a daily basis, the percentage of nonwhites observed to violate speed limits or other traffic rules at that intersection, or some other measure?

Engel and Calnon (forthcoming) report on five different approaches used by researchers to gather baseline data for determining racially disparate outcomes for traffic stops: census data, observations of roadway usage, assessments of traffic-violating behavior, citizen surveys, and internal departmental comparisons. They review various studies that use these strategies to construct baseline measures for traffic stops and describe the strengths and limitations of each.

Census data. Estimates of the driving population derived from decennial census data are commonly used as baseline measures of traffic or pedestrian stops (see, e.g., Harris, 1999b; Zingraff et al., 2000). In practice, the racial composition of stops is often compared with the racial composition of the census population in the immediate vicinity of a stop point, sometimes in combination with motor vehicle records on the racial composition of drivers resident in the area (Engel et al., 2002). However, the flow population can be quite different from the resident population (Zingraff et al., 2000). This is certainly the case with traffic flow: The composition of the drivers passing through a particular neighborhood, particularly on a major highway, may bear little relationship to the neighborhood's residential composition. One might try to take a random sample of drivers passing a particular point (perhaps using pictures taken with a bright flash camera so there will be enough light to permit the identification of race) to establish a distribution across the relevant racial groups.[5] However, there could be serious concerns about the accuracy of such identification, and the sample results could well change with the time of day, day of the week, or season. Pedestrian stops might be more representative of the underlying population but not necessarily so in business districts or high pedestrian traffic areas, where stops are more likely to occur.

More sophisticated—although not necessarily more accurate—estimates of the relevant baseline population have been developed from census data by using the racial composition of neighboring counties weighted inversely by the county's distance from the observation point. Engel and Calnon (forthcoming) suggest using baselines that capture differences in frequency and patterns of driving by race. And estimates for a city with a large minority population have been corrected using census data to take account of the mix using public transportation (Rojek et al., forthcoming), although the validity of such a correction process has not been established. All of these approaches need to be calibrated with observation samples.

Observational data. Reports on racial differences in driving patterns and frequency obtained by observation can be compared with differences in rates of stops, citations, searches, and arrests, although the collection of observational data entails costs that can limit the utility of this method for establishing differential outcomes. Examples of observational studies include those of Lamberth (1994, 1996), using data from rolling surveys of the driving population and traffic violators in New Jersey and Maryland, respectively.

Lamberth (1996) reports on a study in which observers driving at the posted speed limit categorized the racial composition of about 5,700 drivers traveling over the speed limit (violators) or not (nonviolators) on particular stretches of I-95 in Maryland. Lamberth used these data to establish a benchmark of law-violating and law-abiding behavior. Although one can imagine the difficulty involved in spotting the race of drivers in cars speeding past the observers, Lamberth does establish an important point—that

[5]Because census race reports are provided by household members, whereas police stops are based on observation, visual identification of race would need to be compared with self-reports so the census data could be adjusted to reflect the likely distribution that would result from observation.

most of the cars observed (93 percent) were traveling above the posted speed limit, a situation in which police have the ability to stop almost any car for speeding. Lamberth clearly believes that racial differences in stop rates when almost everyone is speeding must reflect racial bias. This conclusion, however, rests implicitly on the proposition that speeding was the only basis for stopping cars on the Maryland highway (although one could look only at those stopped for speeding) and that there was virtually no difference in the distribution of speeds for white and black drivers. Lamberth's own data show that whites were more likely than blacks to be driving at the lawful speed on I-95 in Maryland. (Specifically, 7.9 percent of the white drivers observed in Lamberth's study, but only 3.6 percent of the black drivers, were not speeding.) Indeed, a subsequent study conducted on the New Jersey turnpike using radar devices and cameras to determine car speeds and the race of drivers revealed that blacks did drive at very high speeds more often than whites, which would likely cause them to attract more attention from police.[6]

Yet even if racial differences in the rate of stopping motorists on Maryland highways can be explained by differences in driving behavior, the racial disparities in rates of search for illegal activity conditional on being stopped appear to be quite large. The Maryland State Police reported stopping and searching 823 drivers on I-95 during the observation period of Lamberth's (1996) study; 73 percent of those drivers were black and only 20 percent white (the remaining drivers were other racial minorities). Yet blacks accounted for only 18 percent of the speeding drivers who were eligible to be stopped on I-95 (from Lamberth's data), compared with 73 percent of those who were actually searched (from the police data). Lamberth (1994) obtained similar results in his New Jersey study.

Assessment of traffic-violating behaviors. Few studies have determined whether traffic-violating behaviors vary by race. Lamberth (1994, 1996) tried to establish base rates in his studies; however, he did not determine the severity of violating behaviors. Severity in the case of speeding involves both the rate of speed of a driver and the speed at which state police issue citations, which can differ from state to state. For example, if police in a state routinely allow drivers to exceed the posted speed limit by 10 mph, researchers would need to establish the rates at which different racial groups

[6]The study found that in the southern part of New Jersey, where claims of racial profiling had been most common and where the speed limit was 65 mph, 2.7 percent of black drivers compared with 1.4 percent of white drivers drove faster than 80 mph. The racial disparity was even greater for those driving faster than 90 mph. On the other hand, the study did not find any racial differential in speeding in northern New Jersey areas having speed limits of only 55 mph (Kocieniewski, 2002).

exceed that limit to use in comparisons with stop rates. Engel and Calnon (forthcoming) cite researchers who have estimated the degree to which drivers violate the speed limit (e.g., Lange et al., 2001; Smith et al., 2000) but conclude that their methods still do not fully capture differences in the severity of speeding. One reason is the difficulty of reliably measuring all behaviors associated with traffic-violating behaviors.[7]

Citizen surveys. Researchers may conduct surveys of individuals regarding their driving patterns to create baselines for comparison with data on traffic stops. (They may also conduct surveys of individuals concerning their interactions with police to compare with some baseline.) One advantage of citizen surveys is that they provide self-reports on a driver's race. However, self-reporting is less relevant to race as perceived by the police, who are potentially profiling. Moreover, self-reporting is probably less effective for gathering information on traffic violations because of underreporting by respondents, who may view admitting to such violations as socially undesirable or fail to report their violations for other reasons. Baselines developed from citizen surveys may also be inaccurate as a result of differences in driving patterns across local jurisdictions and in the driving population by day of week or time of day (Farmer, 2001).

Internal departmental comparisons. An alternative to creating external baselines is to use comparisons of rates of stops and other behaviors among police officers to identify typical rates. This method is often used as part of a police department's approach to identifying and studying officers who exhibit problematic behaviors, such as high rates of complaints (Walker, 2001). Walker acknowledges that such an approach would not be effective in departments in which institutional discrimination was practiced (i.e., in which departmental policy, explicitly or implicitly, allowed or encouraged race-based profiling). It would also not be effective in cases in which police data reports did not include officers' names for fear of civil and criminal liability or in which officers manipulated the data in one or more respects (as discussed above).

Summary. Engel and Calnon (forthcoming) conclude that methods for identifying racial disparities in police stops are weak but improving. They suggest the best strategy is to use multiple baseline measures to make comparisons with official police data. To best estimate a baseline population, they suggest using surveys and observational studies conducted in various loca-

[7]It may be that data from jurisdictions that have installed cameras at intersections that automatically take pictures of certain kinds of violations will be helpful in this regard.

tions over a long time period, although such factors as cost and size or composition of geographic areas can impede the collection of appropriate baseline data.

Disparities Versus Discrimination

Assuming that the existence of racially disparate outcomes in law enforcement situations has been established, the second and more difficult analytical challenge is to determine the extent to which race-based profiling explains the measured disparities. Seven of 13 studies of traffic stops conducted between 1996 and 2001 (reviewed in Engel et al., 2002) concluded that racial discrimination by police officers fully explained the observed racial differences in stops (American Civil Liberties Union, 2000; Harris, 1999b; Lamberth, 1996; State of New Jersey v. Pedro Soto, 734 A.2d 350, 1996; Smith and Petrocelli, 2001; Spitzer, 1999; Verniero and Zoubek, 1999).[8] However, these studies have been criticized for not having the right type of data to rule out other explanations for the disparities. For instance, Lamberth's (1996) findings (see above) revealed a disproportionately negative outcome for nonwhites in a population for which the likelihood of being stopped was assumed equal for both whites and nonwhites. Yet it is possible that differences in offense rates existed across these groups and that disparities were in part the result of differences in driver behavior and not police behavior.

The remaining six studies reviewed by Engel et al. (2002) acknowledge that factors other than race, such as differences in driving behavior or in neighborhood characteristics that affect the level of policing, could explain the observed disparities (Cordner et al., 2000; Cox et al., 2001; Lansdowne, 2000; Texas Department of Public Safety, 2000; Washington State Patrol, 2001; Zingraff et al., 2000). For example, Zingraff et al. looked at citation rates for black and white men categorized by age and found an interaction effect between race and age such that blacks did not always have the higher traffic citation rate. Thus, black men aged 22 and younger were 24 percent *less* likely to receive citations than were white men in this age group. (The same was true in comparing young black with young white women.) In contrast, black men aged 23 to 49 were 23 percent *more* likely to receive citations than were comparably aged white men, while black men aged 50 and older were 70 percent more likely to receive citations than their white counterparts.

Generally, Engel et al. (2002) conclude that interpreting the findings from extant studies of racial profiling is problematic because there is no

---

[8]All 13 studies estimated at least some degree of racial disparities in policing behavior.

theory guiding the research and data collection. As we have argued in other areas of analysis of discrimination, such as discrimination in hiring behavior by firms (see Chapter 7), it is essential to have an appropriate model of the process that could lead to racial profiling with clearly articulated and justified assumptions if one is to credit a conclusion about the existence of profiling.

At least two different models could be examined in the area of racial profiling. One model would attribute racial profiling largely to the behavior of individual officers ("bad apples") who are prejudiced against minorities. Another model would attribute racial profiling largely to statistical and institutional discrimination.[9] Each model has implications for data collection and analysis. As in other arenas, the difficulty of causal attribution strongly suggests that multiple approaches and kinds of data should be used to understand the extent and types of racial profiling behavior in law enforcement situations.

PROFILING IN THE CONTEXT OF TERRORISM

Because of renewed interest in the United States in the possible use of profiling to identify and apprehend potential terrorists before they commit violent acts, we briefly examine the challenges of identifying screening factors that could potentially select would-be terrorists with a significantly higher probability than purely random selection. Following the attacks of September 11, 2001, media commentators discussed the possibility of racial or ethnic profiling for selecting airplane flight passengers for additional investigation; some also questioned the value of purely random screening, which results in picking up individuals likely to be harmless (e.g., elderly women) (Quindlen, 2002; Wilson and Higgins, 2002).

We identify two sets of issues for consideration: The first involves the difficulties of specifying an effective profile; the second relates to the possible benefits and costs of profiling—not only monetary costs but also social costs that are difficult to measure yet highly important to take into account. We consider not only racial or ethnic profiling as such but also the use of other profiling factors that correlate highly with race or ethnicity so that minorities are singled out disproportionately when the profile is used (disparate impact discrimination).

[9]As noted above, statistical discrimination occurs when police officers use their belief, for example, that young nonwhite males are more likely to be carrying contraband, to justify targeting this group disproportionately in traffic stops, or rely on data showing higher arrest rates for this group for drug offenses and violent crimes. Institutional discrimination occurs when police departments, overtly or implicitly, condone or encourage racial profiling by officers.

By using such terms as "costs" and "benefits," we do not mean to deny the fundamental importance of the civil rights context in considering the issue of racial or ethnic profiling. In that context, racial profiling is considered statistical discrimination and therefore wrong under any circumstances, whether or not it could be proven that there are costs associated with not profiling. Consider the analogy to free speech. People have a right to express themselves. We do not talk about the benefits and costs of free speech; instead, we say there is a right to free expression that continues to exist even when that free expression poses costs to others. But even that right has limits: It cannot be exercised when the costs to others are very large (e.g., yelling "Fire!" in a crowded theater).[10] Thus, we believe it important to review arguments about effective and ineffective profiles and possible costs and benefits of profiling because arguments for the use of the practice have been and will likely continue to be made in an environment of heightened concerns for public safety.

Developing Effective Profiles

We first review the kinds of additional screening that could potentially help protect the public in such situations as boarding an airplane to provide a context for the possible development of racial or ethnic screening factors. At one extreme, a decision could be made to subject every passenger to intensive scrutiny and interrogation well beyond the previous norm. At this time, however, the public does not appear to be willing to tolerate such a level of scrutiny for all passengers because of the hassles and delays as well as the higher costs for security personnel. Given agreement, however, that some kind of screening is desirable to help prevent a terrorist attack, a procedure must be developed for selecting a subset of passengers to be screened. The selection could be random or, more likely, could be based on several profiling factors. Such factors could include one or more of the following: immutable (or relatively immutable) characteristics such as skin color, sex, and national origin; behavior and dress (e.g., wearing a turban, carrying a backpack, appearing nervous); flight patterns (e.g., purchasing a ticket at the last minute); and background information associated with a

[10]Of course, the analogy is only partially on point. Limiting the freedom of individuals to yell "Fire!" in a crowded theater constrains the freedom of everyone. In contrast, profiling constrains the civil liberties of a subset of persons and leaves the civil liberties of others intact. Hence, although the free speech analogy does suggest that civil liberties are not absolute and have been limited for the public good, it also suggests that their limitation usually imposes constraints that are universally shared. By definition, profiling, to be effective, cannot impose widely shared constraints.

name, address, and date of birth obtained from various databases (e.g., credit card histories).

The goal in developing a screening profile is to identify factors that will select would-be terrorists with a significantly higher probability than purely random selection. Several problems make achieving this goal extremely difficult—in particular, the lack of adequate experience with which to establish the effectiveness of various profiling factors, the ways in which the predictive performance of profiling models can be impaired, and the difficulty involved in setting false-positive and false-negative standards for effectiveness.

Inadequate Data

Data must be available with which to evaluate the predictive power of alternative profiling models in terms of the factors to include and the weight to assign to each factor. In the case of airline security, this evaluation is made most difficult because terrorist incidents in the United States are very rare events, and the estimated numbers of known terrorists and their associates are very small compared with more than 2 million air passengers and the number of innocent people who are profiled on any given day. Even though all 19 of the September 11 attackers were young Middle Eastern men, it is difficult to draw reliable conclusions from this fact regarding the propensity of any other young Middle Eastern men, let alone anyone else, to engage in future terrorist acts, given the many other factors involved and the rarity of terrorist actions.

Even when large numbers of data points are available for analysis, as is true of traffic stops, it is difficult to draw valid conclusions about the relative effectiveness of race or other profiling factors. In this context, effectiveness can be measured by comparing "hit rates" among different groups of automobile drivers—usually defined, for example, as the percentage of drivers whose cars are found to contain contraband (e.g., drugs) among the subset of drivers who are stopped and searched.[11]

Engel and Calnon (2001:Table 1) review 15 studies that examined the effectiveness of racial profiling in traffic, pedestrian, and airport stops. The estimated hit rates (in terms of finding contraband in searches given a stop) varied from under 10 percent to as high as 60 percent. By race, eight studies found similar hit rates in searches for whites and nonwhites, but it is

[11]If the same factor, such as race, is used to determine which drivers to stop and also which of those stopped to search, hit rates could be defined for each group as the percentage of drivers found to be carrying contraband among all drivers stopped.

difficult to interpret these findings lacking other information about the stops. If blacks are stopped and searched at higher rates than whites solely because of racial profiling, similar hit rates may indicate similar propensities for carrying contraband and hence the ineffectiveness of racial profiling.[12] Such a conclusion may not be valid, however, if other factors enter into the profiling.[13]

The remaining seven studies found higher hit rates for blacks and Hispanics compared with whites. Engel and Calnon (2001) conclude that these studies do not provide sufficient evidence about racial or ethnic differences in hit rates either, primarily because of the lack of control for other factors (e.g., extralegal and legal characteristics of the stop) that might influence the likelihood of discovering contraband. Given the undesirability of using racial or ethnic variables in profiling on civil rights grounds, any proposed model for detecting terrorists that includes variables that are highly correlated with ethnicity (and especially ethnic variables) would have to be challenged in terms of their contribution to the predictive value. The model would also have to be evaluated very carefully to determine the reliability of the estimates of each variable's contribution to the model's effectiveness and especially the contribution of those variables directly or indirectly related to ethnicity.

Prediction, Not Causation

A second serious problem with developing effective profiling models is that they are almost by definition predictive, not causal, models. There is no process from which one can infer that such characteristics as wearing torn clothing or a turban or appearing to be of Arab origin are related causally to terrorist behavior; one can only hope to identify factors that have a high correlation with terrorist behavior, which is rare in any case.

In the event a profiling model is developed with factors that are reliably estimated to be highly associated with terrorism at a point in time because causation is not involved, terrorist groups are likely to take steps to invalidate or "game" the profile. Thus, if a terrorist group were able to identify the kinds of characteristics that result in being pulled aside (or not) for additional investigation, it could enlist a person without those characteris-

[12]For estimating hit rates in this situation, the higher stop and search rates for blacks simply provide a larger sample for that group.

[13]If an experiment could be conducted in which people were stopped and searched at random in the same areas in which security personnel initiate stops, it might be possible to examine this issue.

tics to carry out a terrorist act.[14] If this is the case, random screening may be more effective than profiling because it cannot be gamed.[15]

A related problem is that an effective profile would essentially harden the primary targets, which in this case comprise airliners. This effect could cause terrorists to shift their attention to "softer" targets. If so, that would represent success in protecting the primary targets, but it would force attention to the question of how broadly we can protect the wide array of potential targets. Would the same profiling instruments work as well elsewhere (say, on mass transit)? That forces consideration of the broad array of threats and vulnerabilities of all possible targets, an issue that is clearly beyond the scope of this panel.

Standards for Effectiveness

A third problem in developing profiles for such purposes as screening airline passengers is determining the standard by which one judges effectiveness. Because associations are never perfect, any profiling model will fail to detect some terrorists, and models developed with limited data may well generate high rates of false negatives. In other words, such models may fail to select terrorists, especially those who do not fit the profile. Moreover, because the base rate is so low, any profiling model will also generate a very high rate of false positives; that is, it will select many people who fit the profile but are innocent of any crime or criminal intent.

Costs and Benefits of Profiling

The benefits of an effective profiling model are readily stated in general terms. In the terrorism context, they include the possible prevention of terrorist acts that, if not detected, could result in catastrophic loss of lives and property. Furthermore, it might be posited that a high rate of prevention of planned attacks could, over time, discourage terrorist groups from planning further attacks. Because terrorists typically seek to inflict severe damage, societal concern about the potential loss of hundreds or thousands of lives in an attack (as occurred on September 11) is understandably high.

---

[14]For this reason, security agencies strive to keep profiling features secret. The possibility of gaming also argues against using such obvious factors as ethnicity or other features indicative of national origin and turning instead to less obvious factors (e.g., particular travel patterns).

[15]Random screening is not the same as haphazard selection; random screening involves the use of a randomizing device, such as a computer algorithm, to determine which persons to stop.

Yet when an antiterrorism profiling model uses race or ethnicity or factors that correlate highly with race or ethnicity, particularly when such factors are given high weight in the profile, the inevitably large false-positive rates mean that large numbers of members of disadvantaged groups will be falsely singled out for scrutiny. As a result, not only will these individuals experience hassles and delays, they will also likely feel angry, humiliated, and stigmatized. Such stigmatization could well have high negative costs for society at large—if not in the immediate future, then in the longer term. One such cost could be the reinforcement of stereotypes associating minorities with criminal propensities, which could have the damaging effect of reinforcing discriminatory attitudes and behaviors in other domains and having negative feedback for some behaviors of the targets of discrimination (see Chapter 11). A related cost could be the desensitization of the public to the need to be vigilant in protecting important civil liberties, which could lead in turn to readier acceptance of the erosion of civil rights for more and more groups of people who were not originally targeted in profiling. Yet another cost could be possible retaliation (e.g., future terrorist acts) by individuals driven by anger and resentment for being wrongly targeted.

With regard to which groups in society are likely to bear the costs of profiling disproportionately, we note two related points. First, profiling on the basis of race or ethnicity is by its very nature less useful when applied to large groups (when there is only one group, it cannot be used at all). To reduce the false-positive rate, one wants to target profiling on small, narrowly focused groups. The consequence is that the burden of racial profiling will typically fall on smaller groups. Second, such groups may be disadvantaged in other ways and less able to oppose the use of profiling compared with the majority group. Finally, as noted above, it could happen that assessing people on the basis of race or ethnicity in one domain (which is what racial profiling does) may spill over into a reduced concern for civil liberties in other contexts.

Trade-offs

Analysts might consider developing formal cost-effectiveness models to compare the benefits and costs that could be expected from the use of racial or ethnic profiling as a tool in such situations as screening flight passengers to help identify terrorists. Such a task would be challenging in the extreme, although attempts to develop such models could help illuminate the difficult trade-offs involved in assessing the value of profiling. Thus, on the benefit side, it would be difficult and contentious to estimate the number of lives that might be saved through profiling and, further, to estimate the value of those lives. On the cost side, although it might be possible to assign monetary values to the hassles and delays experienced by those law-abiding people who are improperly singled out for scrutiny, it would be very difficult to weigh stigmatization and such larger societal values as the possible serious erosion of civil liberties over the long term.

Ultimately, assessment of the possible use of ethnic profiling in fighting terrorism should involve careful, sober, deliberate consideration by policy makers and the public of three main factors: the desire to protect against the likelihood, albeit very small, of catastrophic terrorist events; the reality that racial (or ethnic or national origin) profiling is likely to be only marginally effective in detecting terrorists in airports and similar venues and, at the same time, will subject many innocent people to harassment and stigmatization; and the importance our society places on protecting core societal values of equal protection and liberties for all.

Over time our society has progressed, through civil war, constitutional amendments, legislation, and court cases, to a conclusion that race-based discrimination in such domains as job markets, housing, and voting is unacceptable and should not be allowed, despite arguments that might be offered to the contrary (e.g., allegations that the presence of disadvantaged racial groups lowers property values). We have reached that conclusion not only for overt race-based discrimination but also for discrimination against racial minorities that results from the use of ostensibly neutral procedures lacking a clear justification. One might argue that similar conclusions extend to discrimination based on ethnicity. Whether our society should maintain that posture in fighting international terrorism is a matter the public might wish to debate. What we have endeavored to do in this brief review is to identify the difficult issues involved, not only in developing profiles but also in assessing their costs and benefits when such vitally important and almost impossible-to-quantify dimensions as public security and core principles of liberty and equality are at stake.

Part III

Data Collection and Research

Part I of our report provided definitions of race and racial discrimination from a social science research perspective and an explication of various types of race-based discrimination and the mechanisms by which overt and subtle forms of discrimination may occur. Part II reviewed the strengths and weaknesses of several broad methods for conducting research on racial discrimination, including laboratory and field experiments, analysis of observational data and natural experiments, and direct measures of racial attitudes and experiences with discrimination from reports in surveys and administrative records.

The discussion of each method in Part II emphasized the difference between descriptive analysis and causal inference. For example, it is one thing to find differences in educational or income levels between minorities and whites and quite another thing to draw a causal inference by which some part of those differences can be validly and reliably attributed to racial discrimination. It is also not straightforward to relate discriminatory attitudes to discriminatory behaviors that have adverse consequences for racial groups. Some of the problems that impair the ability to draw valid causal inferences include that experiments cannot vary the race of any one individual, observational data lack key variables that contribute to differential outcomes among race and ethnic groups, and direct reports of discriminatory behavior and experiences can be biased in one or more respects. In short, there are no ready answers for researchers and policy analysts who are looking to provide definitive information on which to base public and private organization policies about ways to ameliorate discrimination and its effects.

In Part III, we identify priority areas for research and data collection that can help build a stronger base of knowledge about the incidence, causes, and consequences of racial discrimination in a variety of domains. Our discussion emphasizes the need for research that draws on the strengths of different kinds of measurement methods and data sources. Such research requires concerted cooperative efforts among funding agencies that have traditionally funded certain kinds of studies and certain disciplines and among researchers themselves.

Part III comprises Chapters 10–12. Chapter 10 provides a more detailed description than was initially provided in Chapter 2 of federal government standards for collecting data on race and ethnicity and how federal racial categories have changed over time with changing societal conceptions of race. Although not always consistent with scholarly concepts of race, the federal standards are important because they shape much of the data that are available for analysis of racial discrimination, disparities among racial groups, and related topics. The chapter considers measurement issues that affect reporting of race and ethnicity and makes recommendations for continued governmental collection of race data and methodological research to understand reporting effects.

Chapter 11 considers the concept of cumulative discrimination and how racial discrimination may have effects over time and across different domains. Cumulative effects may be missed using some of the methods described earlier in this report. Because so little empirical research has been conducted on cumulative phenomena, either over time or across domains, we treat this topic as a matter of priority for future research. Our discussion in this chapter begins to consider theories and possible approaches that may help researchers interested in studying mechanisms of cumulative discrimination and their effects.

Finally, Chapter 12 provides suggestions to program and research agencies of next steps for building an agenda for research and associated data collection. The aim of this chapter is not to develop a detailed agenda per se; rather, it brings together the recommendations that are in earlier chapters and puts them in a framework of the need for and power of multidisciplinary studies that draw on multiple methods and data sources. Because of the difficulties of measuring racial discrimination, the best analyses will make use of findings from a variety of studies that, ideally, are implemented within a common conceptual and measurement framework.

10

Measurement of Race by the U.S. Government

Since the first U.S. census in 1790, statistics on race have been a prominent part of the nation's censuses and surveys. The Constitution requires the federal government to conduct a census of the country's population every 10 years for use in the allocation of seats in the House of Representatives. Although the uses of the data, the definitions of race, and the methods of data collection have changed, there continues to be intense interest in census data on race and, more recently, on ethnicity (see Anderson, 1988, 2000, for a history of the census).

Today, these data are an integral part of the nation's economic and social policies. Race and ethnicity statistics are used in important and politically sensitive areas, such as the enforcement of civil rights and antidiscrimination laws, and determination of voting districts. For example, state legislatures rely on census race and ethnicity data for geographic areas as small as individual blocks to ensure representation of black and other nonwhite voters within the new boundaries of voting districts that are revised every 10 years. Statistics on race and ethnicity are also used by federal regulators as statistical evidence in employment discrimination lawsuits, as a means of determining whether banks discriminate against minorities when they award home mortgages, and in class action court cases alleging racial discrimination.

To meet these and other information requirements, the U.S. statistical system has changed considerably over the more than 200 years of the country's existence. Nevertheless, problems in the collection of accurate data on race and ethnicity remain. As the population has changed, so have the country's views about defining race. On the one hand, recent news reports have continued to focus on immigration and the country's heritage as a "melting pot" of many races and cultures. On the other hand, prejudice toward disadvantaged racial groups continues to exist, and many members of such groups live in lower economic and social circumstances than the rest of the population. Because the federal government has responsibility for providing information on all groups within the country's population, the statistical system continues to struggle with questions about the number of races for which data are to be collected, how to define and enumerate them accurately, what labels to apply to them, and how to classify persons of multiracial background. In addition, experience has shown the considerable difficulty involved in enumerating the Hispanic population, which appears to bridge both ethnicity and race concepts (see Chapter 2).

This chapter first provides a brief history of the federal government's collection of data on race and ethnicity. It then reviews the standards for government collection of data on race and ethnicity issued by the U.S. Office of Management and Budget (OMB) in 1977 and the revision of those standards in 1997. Next we summarize race and ethnicity data collected in the 2000 census, paying special attention to data for those who selected more than one race. We then discuss some of the issues involved in interpreting and using the new multiple-race data and briefly review research under way in the federal statistical system to resolve those issues. Finally, we make recommendations for continued collection of data on race and ethnicity with categories that are responsive to changing concepts of race among groups in the U.S. population. We further stress the need for sustained research by federal agencies to develop best practices for the measurement of race, to gain knowledge of how different groups report race and of changes in such reporting over time, and to inform users of the meaning of different measures of race and ethnicity.

HISTORY

Article 1, Section 2, of the U.S. Constitution, written in 1787, requires a census every 10 years to determine the number of people living in each of the states. The requirement for data on race grew out of the struggle in the Constitutional Convention over the distribution of power between the North and the South. Because most of the country's slave population lived in southern colonies, the Founding Fathers searched for a way to balance sectional power. The language adopted at the convention—and included in the Constitution—was that "Representatives and direct Taxes shall be apportioned among the several States . . . by adding to the whole Number of free Persons, . . . excluding Indians not taxed, three fifths of all other Persons." The first census in 1790 collected the data required by the Constitution—on free white men (over and under age 16), free white women, and other free persons (who were black)—by direct enumeration or enumerator observation of race. Indians (not taxed) were excluded from the census counts. As time progressed, the labels changed; skin color was frequently introduced to identify racial differences; and in some censuses, more detail about the amount of nonwhite blood was listed—for example, whether Indians were full-blooded and whether blacks were mulatto, quadroon, or octoroon.

After the Civil War, when slavery was outlawed by the Thirteenth Amendment, Section 2 of the Fourteenth Amendment provided for a count of the "whole number of persons in each State, excluding Indians not taxed." Nevertheless, the data system continued to count the number of whites, blacks, and Indians. When large numbers of immigrants from Asia and from Eastern and Southern Europe began coming to the United States, the demand for more information on race, ancestry, ethnicity, and languages rose, and new categories were added. A category for Chinese was added in 1870, and, as more immigrants from East Asia came to this country, other categories (e.g., Japanese, Hindu, and Korean) were added as well. In recognition of the fact that increasing numbers of Hispanic immigrants had established themselves and their families in this country, the 1970 census added a separate question on Hispanic origin.[1] As can be seen from Table 2-1 (see Chapter 2), the number of racial categories continued to grow, and by 1990 there were 15 separate categories plus a separate question with four categories for Hispanic origin (Mexican, Puerto Rican, Cuban, other Hispanic).

In addition, methods of collecting census data changed, with mail increasingly substituting for direct enumeration. By 1960 the Census Bureau began some data collection by mail, and data collection by telephone was used both to conduct entire surveys and to supplement mail collection. These changes obviously made enumerator observation of race impossible. As a result, the manner in which race was determined in government censuses and surveys changed. Today, the household member responding to the questionnaire or survey is asked to identify his or her own race/ethnicity and in some cases that of other members of the household as well.[2]

In the 1960s and 1970s, civil rights laws—such as the Civil Rights Act of 1964 banning discrimination in employment and public accommodations, the Voting Rights Act of 1965, and the Fair Housing Act of 1968—were passed to prohibit the exclusion of disadvantaged racial groups from social and economic privileges. The emergence of the new legislation re-

---

[1]Table 2-1 shows that the 1930 census included Mexicans as a race, but the category was dropped in later censuses. (See Chapter 2 for a fuller explanation.)

[2]The issue of self-identification versus interviewer observation is discussed more fully in Chapter 2.

quired the collection of race and ethnicity data to monitor compliance. Thus, the purpose of racial categorization shifted politically from denying opportunities to disadvantaged racial groups to ensuring compliance with civil rights laws and promoting antidiscrimination policies (Anderson and Fienberg, 1999b; Nobles, 2000). Many nonwhite advocacy groups lobbied for the federal government to continue collecting data on race and ethnicity to ensure civil rights protection for their groups. For example, several Asian American groups insisted that their specific categories be added to the 1980 and 1990 census questions on race.

Government statistical agencies also undertook research on the response effects on race/ethnicity reporting of wording, questionnaire design, data collection techniques, and other aspects of survey design in an effort to obtain better data, expand coverage of minorities, and improve scientific survey methods (Tucker and Harrison, 1995; Tucker and Kojetin, 1996). As it became clear that some groups within the population, especially disadvantaged racial groups, continued to be counted less accurately than others, increasing attention was focused on the problems caused by the differential undercount of these populations and how to overcome those problems.[3]

STANDARDS FOR THE COLLECTION OF RACE AND ETHNICITY DATA

The 1977 OMB Standards

Because of the need for consistent data based on uniform definitions for use in connection with civil rights legislation and monitoring of equal treatment, as well as for other public policy uses of race/ethnicity data, in 1977 OMB developed and issued to federal statistical agencies a set of standards for the collection of such data (Nobles, 2000). Statistical Directive Number 15 established a classification system that included four major categories for race—white, black, Asian or Pacific Islander, and American Indian or Alaskan Native—and two for ethnicity—Hispanic and non-Hispanic.[4] It

[3]For several decades, controversy about the census focused primarily on population coverage and the differential undercount of nonwhite groups. Despite the special steps that have been taken to improve their response rates, more nonwhite than white people have been missed in the census. Because the number of people counted can affect apportionment for the House of Representatives as well as allocations of funds to states and local governments, the undercount issue has been surrounded by political controversy. (See Anderson and Fienberg, 1999b, for a discussion of the issues involved, and National Research Council, 2004, for an evaluation of the problem in the 2000 census.)

[4]The census and other surveys also include various racial subcategories, such as Japanese, Chinese, and Vietnamese.

also encouraged self-identification as the preferred method of collecting data on race. Respondents were instructed to choose only one race. The standards were required to be used in censuses and surveys conducted by the federal government, as well as for federal administrative records and research (U.S. Office of Management and Budget, 1977). Although the same definitions were also used in private surveys (especially those financed by the federal government, such as the General Social Survey), private surveys sometimes collected less detailed data on race, combined categories into broader groups, or formulated the race questions somewhat differently.

By the 1990 census, questions had been raised about the continued relevance of the 1977 standards. Many population changes had occurred since 1977, and the population of disadvantaged racial groups had grown considerably. In fact, the rate of population increase for blacks, American Indians, Eskimos, and Aleuts, as well as for Asians and Pacific Islanders, between 1980 and 1990 had been higher than the rate for the white population. In addition, questions began to be raised about how to enumerate race for children born of interracial unions. Statistical agencies had initiated research on the effects of differences in question wording and placement. They believed research was required on how to define race and ethnicity, which labels to attach to the various categories, and what to do about the rising number of multiracial individuals. The issues addressed in that research were discussed widely with many population groups (e.g., Arabs, Cape Verdeans, Muslim West Asians, and Creoles) who wanted separate categories for population groups not yet included in the census categories and increased detail about countries of origin and languages used. These groups actively campaigned to add their categories to the census. Congressional hearings were held in 1993 (by the House Subcommittee on Census, Statistics, and Postal Personnel), and OMB decided to undertake a complete review of the 1977 standards.

Other kinds of issues were also raised. Many groups, concerned about children of interracial marriages, argued that they should not be forced to select the race of only one of their parents and asked for a new multiracial category. Other groups, worried about the use of racial categories as the basis for antidiscriminatory action, feared that use of a multiracial category would dilute data needed for the nation's civil rights programs (Anderson and Fienberg, 2000; U.S. Office of Management and Budget, 1997).

Research by Federal Statistical Agencies on Race and Ethnicity

OMB established an interagency committee to assist it in its review of the 1977 standards. That committee established an interagency research working group, chaired jointly by the Census Bureau and the Bureau of Labor Statistics (BLS), to develop an agenda for the specific questions to be addressed and the methods to be used for determining how changes might affect the measurement of race and the quality of the data obtained. The working group reviewed the criticisms and suggestions made thus far and developed a research agenda focused in particular on how to enumerate people who identify themselves as multiracial, whether to add new racial categories, whether to change the terminology used for racial categories, and whether to combine race and Hispanic origin in one question or have separate questions (Tucker and Harrison, 1995). In addition, the National Research Council's Committee on National Statistics conducted a workshop on these issues (National Research Council, 1996).

In May 1995, a special supplement to the Current Population Survey (CPS) was undertaken, focused primarily on three issues: (1) the ability for respondents to select a multiracial category; (2) whether Hispanic should be added to the list of races or whether, as in the past, a separate question on ethnicity should be used; and (3) use of such alternative race/ethnicity labels as black, African American, or Negro, and Hispanic, Latino, or Spanish. In the 1996 National Content Survey and the Race and Ethnic Targeted Test, the Census Bureau explored multiracial response options, the combining of Hispanic origin with race, and race wording issues. Other statistical agencies were also involved. For example, the National Center for Education Statistics (NCES) explored how race was recorded in schools, and the National Center for Health Statistics (NCHS) reviewed the determination of race in the administrative records with which it dealt.

The following results were among the most important findings from the CPS supplement study:

- The number of respondents identifying themselves as Hispanic was higher when a separate question on Hispanic origin was followed by another question on race than when Hispanic origin was combined with the race question (see Chapter 2 for discussion).
- In the two test panels (each comprising 15,000 households) in which a question on whether respondents were or were not Hispanic or Latino was followed by a separate question on race, the inclusion of a multiracial category in the race question had little effect on the percentage identifying themselves as Hispanic—10.79 percent reported Hispanic origin when there was no multiracial category in the race question, as compared with 10.41 percent who reported Hispanic origin when the race question included a multiracial category.
- In the two test panels of 15,000 respondents each in which Hispanic, Latino, or "of Spanish origin" was included as a racial category instead of as a separate ethnicity question, smaller percentages reported Hispanic origin—7.5 percent identified themselves as Hispanic when there was no mul-

tiracial category, as compared with 8.6 percent when a multiracial category was included (Tucker and Kojetin, 1996).

- Not surprisingly, the number selecting "other" for race was smaller when the Hispanic question was combined with race.
- The count of the white population was smaller when Hispanic was listed as a category in the race question, apparently because a number of Cubans who identified themselves as white when separate race and ethnicity questions were asked could not do so when Hispanic was listed only as a part of the race question (Tucker and Kojetin, 1996).

Despite these results, it is interesting to note that when asked for their preference, a substantial majority of Hispanics preferred to have the Hispanic question included with race. This was true both for those panels with a separate ethnicity question and for those panels with Hispanic as a racial category, although the former group had a somewhat lower percentage (Tucker and Kojetin, 1996). As mentioned in Chapter 2, different question formats can affect responses to questions on race and ethnic origin. Moreover, there are different perspectives on race and ethnicity even within the Hispanic population, making it difficult to interpret data on race and Hispanic origin from surveys (de la Garza et al., 1992; Denton and Massey, 1989; Harris, 2002).

The BLS test also provided other information. Only a small group (less than 2 percent) identified themselves as multiracial. The fact that respondents could select a multiracial category had little effect on other racial categories, with the American Indian/Alaskan Native group as a possible exception. Although no firm conclusions could be drawn from this test about preference for the use of the term African American rather than black or for Native American rather than American Indian, a sizable minority of each group preferred those terms (Tucker and Kojetin, 1996).

A year later the Census Bureau undertook two surveys to explore some of these issues. The National Content Survey tested the effects of the addition of a multiracial category, placement of the Hispanic origin question, and combinations of those changes. The results of the National Content Survey were similar to those of the CPS supplement. They showed that

- Only about 1.0 percent chose the multiracial category, and the choice had no statistically significant effect on the other racial groups, with the possible exception of the Asian and Pacific Islander category.
- Nonresponse for Hispanics was significantly reduced when the Hispanic origin question came before the race question. Also, this placement increased the number of Hispanics identifying themselves as white in the racial category.

A second Census Bureau test was conducted in the 1996 Race and Ethnic Targeted Test. This test was focused on measuring the effects of changes in the race and ethnic standards, especially on smaller population groups, such as American Indians and Alaskan Natives, Asians, and Hispanic subgroups (e.g., Puerto Ricans and Cubans). It also tested a "mark more than one" format for the race question, finding that the number of respondents reporting Hispanic origin did not decline in the combined race/ethnicity question format compared with the two-question format when respondents could check more than one race. Moreover, response rates were higher for the combined format than for the separate race and ethnicity questions (Hirschman et al., 2000).

In addition, NCES undertook an investigation of how race and ethnic classifications are used in the public schools. It found that 55 percent of all public schools record the race and ethnicity of students only when they first enroll in school, and about one-quarter collect these data each year. Some 45 percent of schools ask parents to select one of the OMB race/ethnic categories for their children, 17 percent ask them to select from a list used by the school district, and in some cases parents may write in their own category. Interestingly, more than one-fifth (22 percent) of the public schools use teacher or administrator observation to determine the race/ethnicity of students. The proportion determined by observation is much higher in the Northeast (44 percent) (National Center for Education Statistics, 1996).

The 1997 OMB Revised Standards

Building on these results, and following public comment and hearings, Statistical Directive Number 15 was revised in 1997 to define the categories to be used in the 2000 census and for other government surveys (U.S. Office of Management and Budget, 1997). The new standards—Standards for Maintaining, Collecting and Presenting Federal Data on Race and Ethnicity—included three major changes. First, five racial categories were to be used in measuring race: black or African American, white, Asian, American Indian and Alaskan Native, and Native Hawaiian or Other Pacific Islander. Second, there was a requirement that respondents be permitted to select more than one race. Third, the question on ethnicity was to be simplified by asking respondents whether they were Hispanic or Latino, and the ethnicity question was to be asked before the race question. Although the standards were to be used by all federal agencies in the future, agencies were permitted to add categories when more detailed data were needed. In issuing the new standards, OMB emphasized that "the categories represent a social-political construct designed for collecting data on the race and ethnicity of broad population groups in this country, and are not anthropologically or scientifically based" (U.S. Office of Management and Budget, 1997:16). The new standards were used in the 2000 census, with the Census Bureau adding the category "other" to the five racial categories established by the standards.

GOVERNMENT DATA ON RACE AND ETHNICITY

The 2000 Census

In the 2000 census, nearly 275 million people or almost 98 percent of the total population identified themselves as one race only, whereas 2.4 percent or 6.8 million people selected two or more races. Of those who chose one race only, 75 percent identified themselves as white and 12.3 percent as black or African American (see Table 10-1). The Asian population, which had grown by 48 percent between 1990 and 2000, was the next largest group identifying with one race (3.6 percent), followed by American Indians and Alaskan Natives (0.9 percent) and Native Hawaiian and other Pacific Islanders (0.1 percent).

These results demonstrate the remarkable increase in the country's racial diversity, both because the white population has not increased as rap-

TABLE 10-1 Race and Hispanic Origin Population in the United States, 2000

Race and Hispanic or Latino Origin	Number	Percent of Total Population
Race		
Total population	281,421,906	100.0
One race	274,595,678	97.6
White	211,460,626	75.1
Black or African American	34,658,190	12.3
American Indian and Alaskan Native	2,475,956	0.9
Asian	10,242,998	3.6
Native Hawaiian and Other Pacific Islander	398,835	0.1
Some other race	15,359,073	5.5
Two or more races	6,826,228	2.4
Hispanic or Latino Origin		
Total population	281,421,906	100.0
Hispanic or Latino	35,305,818	12.5
Not Hispanic or Latino	246,116,088	87.5

SOURCE: Data from U.S. Census Bureau (2001b).

idly as the nonwhite population and because the sizes and mix of many disadvantaged racial groups have changed. For example, 100 years ago, 87 percent of the population was white, 12 percent was black, and only about 1 percent was from some other group. By 2000 the white population had declined to 75 percent; the black population, at 12.3 percent, had risen only slightly as a proportion of those who identified themselves as belonging to a single race; and the Asian population had become a significant racial category (Anderson, 2000).

But whites and blacks do not fully explain the changing race/ethnicity makeup of the American population. The 2000 census counted nearly 15 million people, or 5.5 percent of those identifying with a single race, in the "some other race" category. Further breakdowns by the Census Bureau show that a very large number of those identifying with "some other race" (14.9 million people) were Hispanics, who responded to the race question by selecting the "other" category (U.S. Census Bureau, 2001b). This result suggests that many Hispanics do not identify with the census racial categories and underscores the need for more research on how to measure racial identification more accurately, especially for the Hispanic population. In the 2000 census, the number of Hispanics of any race was close to the number of blacks (including those identified as black only and black together with some other race; see Table 10-1). In the decade between 1990 and 2000, the number of Hispanics in the population increased at a much faster rate than was the case for blacks: The rate of increase was 57.9 percent for Hispanics, 3.5 times the 15.6 percent rate for blacks.

The multiracial population counted in the 2000 census was small—only 6.8 million people or 2.4 percent of the total population. Those identified as belonging to more than one race in response to the race question were young (4 percent of the population under age 18); only about 4 million (1.9 percent of the population over age 18) were adults (U.S. Census Bureau, 2001b). The group included many Hispanics; in fact, 2.2 million Hispanics selected more than one race—nearly one-third of the 6.8 million who reported two or more races. This finding suggests once again that Hispanics have differing conceptions of race and ethnicity and are not certain how to respond to the racial categories on the census questionnaire (see Chapter 2). Categories checked by the multiracial population varied, with such combinations as white and American Indian, white and Asian, white and black, and white and other.

As part of its work on the new American Community Survey, planned to replace the decennial census long form in 2010, the Census Bureau fielded a large (700,000 household) survey in 2000 to provide data for comparison and analysis with the 2000 census long form. Estimates for race in the new survey, called the C2SS, differed in some ways from those in the census,

TABLE 10-2 Household Data on Race and Ethnicity in Census 2000 and C2SS

Race/Hispanic Origin	Census 2000	C2SS	C2SS–Census
Total	273,643,273	273,643,269	–4
White alone	206,127,572	211,867,275	5,739,703
Black or African American alone	32,939,206	32,256,169	–683,037
American Indian and Alaskan Native alone	2,400,916	2,117,034	–283,882
Asian alone	10,037,229	10,453,603	416,374
Native Hawaiian and other Pacific Islander alone	388,153	436,612	48,459
Some other race alone	15,053,131	10,700,143	–4,352,988
Some other race Hispanic	14,600,195	10,107,129	–4,493,066
Two or more races	6,697,066	5,812,433	–884,633
Two or more races/Hispanic	2,181,583	1,643,812	–537,771

NOTES: Numbers for both the census and the C2SS are for persons in households. The last column is the difference between C2SS and the 2000 Census (i.e., C2SS minus Census). The CS22 did not cover group quarters.

SOURCE: Unpublished data from the Racial Statistics Branch, U.S. Census Bureau.

even when corrected for differences in coverage (see Table 10-2).[5] The C2SS found slightly more whites and Native Hawaiian and other Pacific Islanders than did the census and somewhat smaller estimates of multiple races as well as for the "other race" category. Those identifying themselves as being of two or more races amounted to 2.1 and 2.4 percent of the household population in the C2SS and the census, respectively. Those reporting "some other race" were a smaller group—amounting to 3.9 percent in the C2SS versus 5.5 percent in the census. Once again, Hispanics appear to account for most of the difference, as the C2SS showed more white Hispanics and fewer "some other race" Hispanics than did the census. Unfortunately, the questionnaires were designed by different groups, and the layout of the race questions was somewhat different in the two surveys, with the C2SS race questions being printed horizontally across the questionnaire and the cen-

[5]This survey was designed to cover a large sample of households. It excluded group quarters (e.g., people living in nursing homes and in prisons and students living away at school). Because the census is intended to cover all people in the country, it includes those living in group quarters. Table 10-2 therefore includes separate data for persons in households so that the results of the C2SS can be compared with those of the 2000 census. It should be noted that the C2SS, as a sample survey, is subject to sampling error.

sus questions printed vertically. Census Bureau staff believe this difference in layout may account for much of the difference in the results between the two surveys.

Race in Other U.S. Government Surveys

The wording of race and ethnicity questions and their placement on questionnaires have often differed among government surveys. Self-enumeration is used to the extent possible, but that does not mean each person covered in a survey always responds for himself or herself. In many of the major government surveys, the questions are asked of only one person, the reference person, who responds to all questions for all members of the household. The agencies currently are working to implement the revised race and ethnicity standard so that all future surveys will include a question about Hispanic/Latino origin before the race question is asked, and all respondents will be permitted to select more than one race. The agencies anticipate that all major surveys will have complied with the new standard by 2003, if not before.

One federal government survey, the Health Interview Survey (HIS), collected by the Census Bureau for NCHS, has a 20-year history of asking respondents to select one or more racial categories. In addition, those who select multiple racial categories in the HIS are asked to indicate which of those races "would best represent your race." These data will be especially useful to the statistical system for understanding issues of data presentation and development of multiracial historical series.

Ongoing Research

The Census Bureau and BLS continue to plan and carry out research designed to study issues associated with the collection of data on race and ethnicity. In particular, they plan to conduct research on racial identification of children in surveys and on potential effects of the mode of data collection on responses on race.

In May 2002 BLS tested the new race question in a supplement to the CPS. The question wording varied by age—information for household members aged 12 and older was obtained about the race the person considered himself or herself to be, while information for younger household members was obtained about the race the respondent considered the child to be. The new race question, with the wording variation by age, was used in the CPS beginning in January 2003.

In addition, the Census Bureau and BLS will conduct field tests in an attempt to determine the effect of different modes of data collection on race responses. These tests will include collection by computer-assisted telephone interviewing, computer-assisted personal interviewing, and personal enumeration by use of paper and pencil. The two agencies have agreed on the following wording for the race question on the tests: "Please choose one or more races that [you/name] consider[s] [yourself/himself/herself] to be: White, Black or African American, American Indian or Alaskan Native, Asian, or Native Hawaiian or Other Pacific Islander." Cooperative efforts are also under way among the statistical agencies to agree on the wording of race and ethnicity questions to achieve uniformity across most of the government surveys.

ISSUES IN THE REPORTING OF DATA ON MULTIPLE RACES

Because of the new OMB standard on enumerating the multiracial population, government statistical agencies must address a number of issues. There may be important reasons for some surveys to use differing approaches, but many of the same issues must nonetheless be addressed. For example, how and how often should multiple races be included in ongoing releases and other publications? How can confidentiality be maintained when samples of those selecting particular multiracial categories are small? How can the agencies ensure that all of the multiracial categories that are published are statistically reliable? How should the new multiracial data be linked to the old single-race data for purposes of historical analysis? How should the data be mapped across various sources, and how will data users identify mismatched data? Finally, for agencies that rely on administrative records data for some or all of their data, how can consistency between the survey and the administrative data be maintained? These are all important questions, and many statisticians and analysts within the federal statistical agencies have been examining alternative approaches to addressing them. Many are still under study, and much will depend on the purpose of the analysis to be undertaken (Tucker et al., 2000).[6]

In preparation for a review of statistical agency action on the new racial guidelines, representatives of seven government agencies held discussions with one another: the Census Bureau, BLS, NCHS, NCES, the Bureau of Justice Statistics (BJS), the National Science Foundation, and the statistical policy group at OMB. These discussions revealed that considerable progress has already been made toward implementing the new guidelines in the government's surveys, although several problem areas remain. A brief dis-

[6]Tucker et al. focus on alternative methods of linking the new race data to data collected under the old guidelines. They describe a number of alternative approaches and conclude that "it is likely that which method is best at matching a reference distribution for outcome measures will depend on the outcome being examined" (2000:21).

cussion of the most important of these follows, along with some proposed solutions.

Publication and Release of Data

Recognizing the problems inherent in dealing with the 63 racial categories that can be developed from information in the 2000 census, OMB, in its discussion of the 1997 standards, identified the need for further research by federal statistical agencies on methods for reporting the numbers of people who selected more than one race (U.S. Office of Management and Budget, 1997). In addition to developing publication rules for multiracial data from their own surveys, the federal statistical agencies needed to ensure that data required for enforcement and monitoring of civil rights could be made available. While the statistical agencies were engaged in the research necessary to develop rules for the publication of multiracial data for their own publications, OMB issued Bulletin Number 00-02 as guidance to the government's executive branch on the aggregation and allocation of racial data for civil rights monitoring and enforcement (U.S. Office of Management and Budget, 2000).

Rules for Combining Multiracial Data for Civil Rights Cases

OMB Bulletin 00-02 lists the five single-race categories and four additional multiracial categories—American Indian or Alaskan Native and white, Asian and white, black or African American and white, and American Indian or Alaskan Native and black or African American. The guidance establishes aggregation rules for agencies to determine counts of multiracial groups, providing for the "collection of information on any multiple race combinations that comprise more than one percent of the population of interest," one example being that, "in Hawaii, there may well be combinations of racial groups that meet this threshold such as Native Hawaiian or Other Pacific Islander, and Asian" (U.S. Office of Management and Budget, 2000:1-2).

Moreover, allocation rules for civil rights monitoring and enforcement provide that, at the aggregate level, multiracial responses combining a nonwhite race with white are to be allocated to the nonwhite race for analysis purposes. Responses that include two or more nonwhite races are allocated to the nonwhite race on which the alleged discriminatory behavior was based. When action requires assessing disparate impact discriminatory patterns (see Chapter 3), the patterns are to be analyzed "based on alternative allocations to each of the minority groups" (U.S. Office of Management and Budget, 2000:2). These guidelines have been criticized by civil rights advocates and racial theorists as arbitrary rules for collecting and tabulating racial data (see Harris, 2002; Harrison, 2002; see also Chapter 2).

Publication of Multiracial Survey Data

The research conducted by the government's statistical agencies has focused on rules for deciding how much multiracial data to publish. In the past, when sample sizes permitted, most agencies published data on whites, blacks, Hispanics, and a category of "other races." In the future, in accordance with the new OMB standards, efforts will be made to publish data separately for Asians, a category that increased in size in the 2000 census. The expectation is that data for the Hispanic population will be improved considerably once the change in placement of the Hispanic/Latino question has been fully implemented in government surveys because the new procedure should result in better coverage of that population. For example, the CPS research mentioned above showed that use of a separate Hispanic origin question placed before the race question significantly reduced Hispanic nonresponse (Tucker and Kojetin, 1996).

However, most of the agencies still have not decided how or how often to publish data for the multiracial categories. Because the multiracial population thus far appears to be quite small, sample sizes for most household surveys will make monthly publication consistent with confidentiality rules difficult if not impossible. The probability is that those who select multiple races in the surveys will most often be combined with those placed in the category for "other races" or included in the total. This will surely be the case for regular monthly publication of data from such surveys as the CPS and the HIS. The agencies will publish data for the multiracial category separately when tests show that the numbers involved are sufficiently large to make the data reasonably accurate. For most of the important, large household surveys, therefore, multirace data will be published at best on a quarterly basis and in some cases only on a semiannual or annual basis. Some agencies, such as NCHS, have indicated that they will attempt to make multiracial data available by pooling the data over several years because confidentiality restrictions will make it impossible to publish them more frequently. BJS believes that samples of criminal events collected in the Crime Victimization Survey are far too small to warrant separate publication of multiracial data, although the bureau intends to collect such data. It is unlikely, therefore, that the socioeconomic aspects of the multiracial categories will achieve much prominence for some time to come.[7]

[7]Each of these surveys has a different sample design and somewhat different use of interviewers and telephone responses. The methods employed in the collection of data also differ, because the government uses a variety of techniques—paper-and-pencil personal interview collection, computer-assisted personal interview collection, and computer-assisted telephone collection. For repetitive, time-series surveys, the household person responding to the survey also can differ at various times, depending on the subject matter and which household members are available for the interview. Each of these processes, as well as others, can affect the quality and consistency of the data obtained.

Time-Series Data

Research is under way in several of the statistical agencies on methods for developing historical series using the old and new racial categories. The alternative—to announce a break in the series—is unattractive, especially for those who analyze trends. BLS and NCHS have undertaken research to work the new data backwards so that they can be interpreted as a single series, before and after 2000.

Several data sources are being used in this research to bridge the new and old racial categories. BLS has arranged for all respondents to the National Longitudinal Survey to be asked the race question a second time in order to obtain data for the same people using the old and new racial categories. BLS has also undertaken research to use information collected in special supplements to the CPS to introduce a CPS historical series in 2003, using population weights for both 1990 and 2000.

Many of the agencies will also make use of the time series developed in the HIS to help develop a bridge to the old racial categories. For many years, the HIS has permitted respondents to select one or more races, and for those who do so, ask a follow-up question to determine the race with which the respondent identifies most closely. These data should provide a reasonable foundation for developing historical data for the NCHS health surveys and may also assist other agencies in linking data reported under the 1977 and 1997 standards

Administrative Data

Although education and employment data may be available as early as 2005, it appears unlikely that data collected from other forms and administrative records will provide information on multiple races in the near future. BJS, for example, works with probation offices, jails, and state correction agencies to collect data from their records, but in most cases these forms include very limited racial data. In the case of the vital records system developed through cooperation between NCHS and the states, the problem is that data on race either are not present at all or are subject to considerable understatement. For example, the race of a child is not recorded on the birth certificates in most states. Information on race on death certificates is usually furnished by physicians or funeral directors, who may have little knowledge of the deceased. The result is that racial information on death certificates may be inaccurate or not reported at all.

Population Controls

Census population estimates, together with up-to-date data on immigration, emigration, and births and deaths, are used as controls for all government sample surveys to weight the sample data to totals that represent the population groups the sample has been selected to represent. It is clear that much of the success in the handling of the post-2000 census data on race and ethnicity will depend on the manner in which the new population controls are developed. If there are coverage or estimation errors in the population weights, the data from the surveys will reflect those errors. Methods for developing population counts are especially complex because the Census Bureau must develop these controls not just for the country as a whole but also for states and, for some surveys, for a number of individual areas, some of them quite small. The quality of these population counts—and the detail in which they are developed—can affect the presentation of data from all of the nation's household surveys, as well as private research that rely on these data.

SUMMARY, CONCLUSION, AND RECOMMENDATIONS

Data constitute an important tool in defining discrimination and in assisting in the reduction of inequities in treatment based on race. This is especially true of statistical information on race and ethnic categories. Although the country has been collecting such information for more than 200 years, and scientific advances have greatly improved the data collected by the federal government, race and ethnicity data remain difficult to define, and racial categories are frequently not well understood.

The federal government's collection of data on race has changed over time, in part reflecting changing conceptions of race in the United States. In 1997, OMB revised standards for the collection of data on race and ethnicity in the 2000 census and other government surveys. The changes resulted in more realistic categories and labels and permitted respondents to select more than one race. Although government standards are not always consistent with or comparable to scholarly discussions of the meaning of race, the collection of such data is useful.

> **Conclusion:** *Data on race and ethnicity are necessary for monitoring and understanding evolving differences and trends in outcomes among groups in the U.S. population.*

Differences in ancestry, language, and culture, as well as societal attitudes toward race and ethnic differences, influence how people identify with race. The growing number of Hispanics and individuals who identify with more than one race adds to the complexity of measuring racial self-identification. In addition, different respondents interpret questions differently, which can affect the accuracy of their responses (e.g., Hispanics have differing conceptions of race and ethnicity).

Because the nation's statistical system is highly decentralized, the data produced by the system may lack consistency and vary considerably in quality. We lack information about such differences among racial subgroups. In addition, survey practices and data collection methods differ considerably depending on the type of survey and the kind of respondent. Federal government guidelines make self-identification of race by respondents the preferred means of collection, "except in instances where observer identification is more practical, e.g., completing a death certificate" (U.S. Office of Management and Budget, 1997:8). Even so, in many federal government surveys—including the census, from which the country's most comprehensive data on race and ethnicity are developed—the household member who responds to the questions identifies the race and ethnicity of all members of the household.

In many respects, the changes in race/ethnicity categories incorporated into the 2000 census are useful. During the next decade, the federal government needs to further improve race and ethnicity data.

Recommendation 10.1. The federal government and, as appropriate, state and local governments should continue to collect data on race and ethnicity. Federal standards for race categories should be responsive to changing concepts of groups in the U.S. population. Any resulting modifications to the standards should be implemented in ways that facilitate comparisons over time to the extent possible.

Recommendation 10.2. Data collectors, researchers, and others should be cognizant of the effects of measurement methods on reporting of race and ethnicity, which may affect the comparability of data for analysis:

- **To facilitate understanding of reporting effects and to develop good measurement practices for data on race, federal agencies should seek ways to test the effects of such factors as data collection mode (e.g., telephone, personal interview), location (e.g., home, workplace), respondent (e.g., self, parent, employer, teacher), and question wording and ordering. Agencies should also collect and analyze longitudinal data to measure how reported perceptions of racial identification change over time for different groups (e.g., Hispanics and those of mixed race).**

- **Because measurement of race can vary with the method used, reports on race should to the extent practical use multiple measurement methods and assess the variation in results across the methods.**

11

Cumulative Disadvantage and Racial Discrimination

In earlier chapters, we reviewed various methods for measuring certain types of racial discrimination, including laboratory and field-based experiments (such as audit studies), statistical inference methods for observational data, and surveys of racial attitudes and experiences of discrimination. Analysts typically use these methods to identify and measure discrimination that occurs at a certain point in time within a specific domain. In this chapter, we observe that important effects of prior discrimination may be missed with these methods. The discussion expands the potential impact of racial discrimination to include cumulative effects over time, as well as the interaction between effects of discrimination experienced in one domain and at one point in time and events that occur in other domains and at other points in time.

Our concern here is with effects that operate over time. For instance, studies might measure small effects of discrimination at each stage in a domain (e.g., hiring, evaluation, promotion, and wage setting in the labor market), thus leading one to conclude that discrimination is relatively unimportant because the effects at any point in time are small. Over time, however, small effects could cumulate into substantial differences. We identify three primary ways through which discrimination might cumulate:

- *Across generations.* Discrimination in one generation that negatively affects health, economic opportunity, or wealth accumulation for a particular group may diminish opportunities for later generations. For instance, parents' poor health or employment status may limit their ability to monitor or support their child's education, which in turn may lower the child's educational success and, subsequently, his or her socioeconomic success as an adult.

- *Across processes within a domain.* Within a domain (e.g., housing, the labor market, health care, criminal justice, education), discrimination at an earlier stage may affect later outcomes. For instance, discrimination in elementary school may negatively affect outcomes in secondary school and diminish opportunities to attend college. Even single instances of discrimination at a key decision point can have long-term cumulative effects. For example, discriminatory behavior in teacher evaluations of racially disadvantaged students in early elementary school may increase the probability of future discrimination in class assignments or tracking in middle school. Similarly, in the labor market, discrimination in hiring or performance evaluations may affect outcomes (and even reinforce discrimination) in promotions and wage growth.
- *Across domains.* Discrimination in one domain may diminish opportunities in other domains. For example, families that live in segregated neighborhoods may have limited access to adequate employment and health care.

This chapter is necessarily quite speculative. Very little research has attempted to model or estimate cumulative effects. In part, this is because modeling and estimating dynamic processes that occur over time can be extremely difficult. The difficulty is particularly great if one is trying to estimate causal effects over time. That is, we are ideally interested in measuring the presence and effects of racial discrimination at multiple points in a dynamic process.

Chapters 6 and 7 address the difficulties involved in credibly measuring the presence and effects of racial discrimination within one domain at a point in time, including the difficulty of estimating how discriminatory behavior contributes to a difference in observed outcomes. Measuring the impact of discrimination on outcomes over time is even harder. Although some research attempts to track cumulative disadvantage, there is a paucity of studies that credibly measure an effect of discrimination and trace its causal effects over time.

Because the cumulative question has rarely been discussed, this chapter begins by fleshing out the concept of cumulative effects of discrimination that we first introduced in Chapter 3. We then provide a more detailed discussion of the three avenues listed above through which cumulative discrimination may occur (across generations, across processes within a domain over time, or across domains over time). Next, we briefly describe three existing approaches (in three distinct literatures) to modeling the dynamic processes of cumulative disadvantage and discrimination. Finally, we turn to issues involved in trying to measure the magnitude and importance of cumulative disadvantage and trace out the effects of racial discrimination over time. We sketch several possible approaches while commenting on the difficulties involved in their implementation. This measurement discussion is best viewed as describing a possible future research agenda; there has not been enough work in this area for us to make statements about which approaches are most promising or persuasive.[1]

THE CONCEPT OF CUMULATIVE DISCRIMINATION

We briefly elaborate on the concept of cumulative discrimination and how it relates to other concepts and measures, making four main points. First, *by cumulative discrimination we mean a dynamic concept that captures systematic processes occurring over time and across domains.* Discrimination has cumulative effects when a discriminatory incident affects not only the immediate outcome but also future outcomes in one's own lifetime or in later generations. For example, slavery or racial exclusion of certain groups in the past that limited occupational earnings may have negatively affected wealth accumulation for future generations among these groups (Sacerdote, 2002).

One particularly interesting aspect of the dynamic processes that may generate cumulative discriminatory effects is the possibility of feedback effects (Blau et al., 1998). That is, cumulative discrimination may be more than an additive process in which the effects of discriminatory incidents sum over time to form larger and larger outcome disparities. The probability of future discriminatory events may be causally related to past discriminatory events, so that current discrimination may increase the probability of future discrimination. For example, in the education system, any bias in teachers' expectations about the academic performance of black or Hispanic elementary school students may negatively influence the students' performance (e.g., by generating self-fulfilling prophecies) (Jussim, 1989, 1991; Jussim and Eccles, 1992; Rosenthal, 2002). Over time, lower performance by such students may do the following: reinforce negative stereotypes; influence teachers' expectations about the performance of students from these groups, resulting in even poorer performance by them (see Ferguson, 1998); and lead to their experiencing greater discrimination later in life. In an example from the labor market, discrimination in job hiring could make individuals in the target group reluctant to invest in future education or training, permanently lowering their skill levels. This outcome could in turn reinforce employer prejudices and lead to ongoing hiring discrimination in the future.

---

[1]At points in this chapter, we reference suggestions from various colleagues to whom we wrote, seeking their advice about research on cumulative discrimination.

Second, *measures of discrimination that focus on episodic discrimination at a particular place and point in time may provide very limited information on the effect of dynamic, cumulative discrimination.* For example, very small amounts of bias at each level of a multilayer organization can result over time in major bias at the top level with regard to the composition of top management (Martell et al., 1996). Similarly, the amount of discrimination measured at any one stage in a particular domain may be relatively small (e.g., racial steering of housing applicants), yet small effects cumulating over individuals' lifetimes may yield large disparities (e.g., residential segregation). Williams and Neighbors (2001) posit that examining a single instance of discrimination may result in substantially understating the overall level of discrimination. For instance, chronic, everyday exposure to small amounts of discrimination may occur in school, at work, or in public settings. Exposure to chronic discrimination can negatively affect outcomes across multiple domains throughout an individual's life course.

Third, *current legal standards do not adequately address issues of cumulative discrimination.* In the legal sense, discrimination is conceived of as an event that happens at a specific time and place, rather than as an ongoing process yielding cumulative disadvantage over time. Standards of disparate treatment and disparate impact typically focus only on the current environment and give little weight to prior discriminatory behaviors and practices that affected earlier generations, other domains, or past experiences. Therefore, the concept of cumulative discrimination is not addressed directly by current legal definitions of or legal remedies for discrimination. The greater the extent and burden of cumulative discrimination, the more powerful are the arguments for broadly tailored remedies (legal or legislative) that address large racial disparities, rather than narrowly tailored legal remedies that address specific instances of discrimination.

Fourth, *the effects of cumulative discrimination can be transmitted through the organizational and social structures of a society.* While individual discriminatory behaviors can certainly have cumulative effects, the ways in which discriminatory effects are "transmitted" across domains and over generations often depend on social organization. For instance, policies and processes that produce inequalities in housing and labor markets (e.g., segregated neighborhoods and occupations) can also produce inequalities in education (e.g., segregated schools with fewer resources) (see Mickelson, 2003). Faced with persistent discrimination and societal disadvantage, disadvantaged racial groups may make life choices under these racially biased conditions that limit their life chances and future opportunities. Hence, any discussion of cumulative discrimination will move us to closer consideration of the institutional and social processes through which disadvantage is transmitted.

Although there is a paucity of empirical work attempting to measure the cumulative effects of discriminatory events or to determine the extent to which past discrimination causes present disadvantage, the large and continuing racial disparities in the United States are at least consistent with the possibility that cumulative discrimination is important. In this chapter, our goal is to consider possible approaches to identifying and measuring the cumulative effects of discrimination.

AVENUES THROUGH WHICH CUMULATIVE DISCRIMINATION MAY OCCUR

Cumulative Discrimination Across Generations

Discriminatory effects can cumulate over lifetimes and across many generations; that is, discrimination against parents in one generation may directly affect outcomes for their children and indirectly affect life opportunities for subsequent generations (e.g., through poorer education or poorer health). Few studies are able to link discrimination experienced by parents directly to children's outcomes, but research has suggested a variety of channels through which such a link may occur. For instance, continued racial segregation in housing has ongoing implications for wealth levels and accumulation in future generations (Conley, 1999; Oliver and Shapiro, 1995). Several researchers have found that parents' education can influence youths' educational aspirations and attainment (Duncan and Magnuson, 2001; Mare, 1995; U.S. Department of Education, 2001b). Moreover, knowledge about and expectations of going to college influence not only this generation's college attendance but also the knowledge and expectations of the next generation (Massey et al., 2003). Thus, parents who experience discrimination may socialize their children to avoid certain places or situations, or they may have educational and occupational experiences, knowledge, or goals that limit prospects for their children (see Bowman and Howard, 1985; Boykin and Toms, 1985; Hughes and Chen, 1999).

Discrimination against parents at one point in time may limit prospects for their children even if the discriminatory behavior comes to an end or the children face no discrimination. Although evidence of the impact of parental income on child outcomes is mixed, recent work suggests that parental income may be particularly important for younger children in low-income families (see Duncan and Magnuson, 2002, for a summary). For example, if parents cannot afford to live in better school districts or provide extracurricular learning opportunities, their children are likely to do worse in school. Thus, factors, including discrimination faced by parents, that limit parental income may lead to lower achievement by their children.

An ongoing debate within sociology and other disciplines concerns the extent to which outcomes for one generation persist over time and spill over into subsequent generations (see Alba, 1990; Farley, 1990). In particular, some suggest that racial and ethnic differentials narrow and even disappear after one or two generations (Gordon, 1964; Park, 1950). Others argue that differentials persist across generations, affecting human capital accumulation (Alba et al., 2001; Borjas, 1994). Borjas finds that education and skill differentials between immigrant and native U.S. workers (based on wage data from the 1910, 1940, and 1980 censuses) are important determinants of the education and skills of their children and grandchildren. He also shows that differentials converge after four generations; however, experiences among different immigrant groups are qualitatively different and should not be generalized.[2] Sacerdote (2002) finds convergence in outcomes (literacy and occupation) between descendants of U.S. slaves born in the nineteenth century and descendants of free blacks within two generations after the end of the Civil War. Thus, after slavery ended, former slaves caught up to free blacks, and the large literacy gap that existed between them disappeared.[3]

Discrimination Across Processes Within a Domain

As individuals engage in sequential interactions in the labor or housing markets or within the health care, criminal justice, or education systems, discriminatory experiences may have cumulative effects. For instance, discrimination early in one's career may affect performance evaluations, promotions, and wages. Weinberger and Joy (2003) indicate that wage gaps are small between college-educated blacks and whites when they are first hired, but the gaps increase in the years after they leave college. This finding is at least consistent with a theory of cumulative discrimination (although there may be other explanations as well). In education, as noted above, biases in teacher expectations in the early years of schooling may affect later educational experiences and student performance (Ferguson, 1998; Jussim, 1989; Jussim et al., 1996; Murray and Jackson, 1982–1983). Ferguson, for instance, concludes that teachers' perceptions and expectations, which may build sequentially over time from kindergarten through

[2]Although differences (e.g., in literacy rates) were relatively low among different immigrant groups in the early twentieth century, European immigrants who were assimilated into U.S. society overcame many institutional and cultural barriers that non-European immigrants (e.g., Mexicans) did not (Alba et al., 2001). Thus, there may be slower convergence of differentials over time between non-European immigrants and U.S. natives.

[3]Sacerdote did not examine black–white differences but assumed there were fewer cultural and institutional barriers between slaves and free blacks than there were between blacks and whites at the time.

high school, probably contribute to black–white differences in educational achievement. Similar examples can be seen in cumulative interactions within the criminal justice or health care systems.

Single instances of discrimination that affect key outcomes may have cumulative effects even if no future discrimination is experienced. Even more problematic, discriminatory effects at one point in time may place an individual at greater risk of future discrimination, leading to even larger cumulative effects. The institutional processes that evaluate individuals and determine their progress through a system over time can be important in transmitting cumulative discriminatory effects. For instance, most schools use tracking—that is, grouping students into classes or special programs by achievement level. This process typically begins in elementary school and continues through secondary school (Alexander et al., 1999; Kornhaber, 1997; National Research Council, 1999). Several researchers have shown that track divergence occurs over time (Gamoran and Mare, 1989; Kerckhoff, 1986). Mickelson (2003) determined that racially disadvantaged students (e.g., blacks, Hispanics, and Native Americans) are found disproportionately in lower educational tracks for which curricula and instructional practices are weak (see also Hallinan, 1998; Lucas, 1999; Lucas and Berends, 2002; Mickelson, 2001; Oakes, 1985, 1994; Oakes et al., 2000; Welner, 2001; for a more extensive discussion and references, see Mickelson, 2003).

Mickelson (2001) conducted a survey of all middle and high schools in the Charlotte–Mecklenburg school district, long considered a model desegregated district. An examination of all eighth-grade middle school English placements showed that of those who scored in the highest decile as second-grade students, whites were about four times more likely to be in the highest track compared with their black counterparts. This disparity was evident even after controlling for prior achievement, family background, and other factors. Mickelson (2003) concludes that systematic track placements that differ because educators teach, advise, or schedule blacks differently than whites constitute evidence that discrimination is occurring.

Discrimination Across Domains

Discrimination in one domain may also affect outcomes in other domains. In education, discrimination may negatively affect later academic achievement, which in turn may limit access to employment opportunities and affordable housing. Discrimination in hiring can affect residential options, which can also affect schooling and employment options. Discrimination in housing markets is particularly problematic because the distribution of housing affects factors associated with place of residence, such as education, access to jobs, and home equity. Yinger (1995) estimates that housing discrimination lowers the total net worth of black households by $1,335 billion and of Hispanic households by $600 billion.

Past findings on the influence of neighborhood characteristics on other domains are mixed (Jencks and Mayer, 1990). Some of the most persuasive research has occurred in recent years, as the U.S. Department of Housing and Urban Development has funded a series of randomized experiments seeking to identify the effects of residential location on family and child outcomes. The Moving to Opportunity studies are following families who volunteered for relocation out of public housing projects. A randomly assigned subset of these families received help in relocating to low-poverty neighborhoods only (with ongoing rental subsidies through Section 8 vouchers). Results to date indicate that families who moved to low-poverty neighborhoods, compared with the comparison group, have experienced higher employment rates and income, better housing conditions, less exposure to criminal activity and violence, and improved physical and mental health among adults and children (Del Conte and Kling, 2001; Ludwig et al., 2001). The results vary somewhat across different cities, but they are consistent with a review of related (nonexperimental) research by Leventhal and Brooks-Gunn (2000). Many argue that racial discrimination has been highly important in determining residential location patterns (Massey and Denton, 1993). The Moving to Opportunity studies indicate how residential location can have substantial effects on other outcomes.

There is additional research linking residential location with outcomes in other domains. For instance, the so-called spatial mismatch literature investigates how residential location may influence job finding and unemployment (Kain, 1968). Recent work suggests that spatial mismatch results in poor access to jobs, longer commutes, lower wages, and lower employment for low-skilled nonwhite workers (Ihlanfeldt and Sjoquist, 1998; Mouw, 2000). Although these findings suggest that the housing market affects labor market outcomes, studies of firm relocation indicate how exogenous changes in the labor market also affect residential location and housing (Fernandez, 1997; Zax, 1989).

Discrimination in the criminal justice system may affect various other outcomes for disadvantaged racial groups as well. Few studies make the link to discrimination, but existing research does indicate how discrimination at one stage could influence outcomes at another. Compared with whites, blacks and other disadvantaged groups are much more likely to be sent to prison and sentenced to longer periods of incarceration (Tonry, 1996). High rates of black incarceration can disrupt schooling, leading to poor employment prospects and job instability (Sampson and Laub, 1997; Western, 2002; Western and Pettit, 2002). Lochner (1999) argues that education, employment, and crime are all causally linked, so discrimination in any one area will affect other areas.

Disparities in incarceration rates also have a negative impact on the health of disadvantaged racial groups. Weich and Angulo (2002) note that prison overcrowding and lack of health care led to an outbreak of tuberculosis in the early 1990s. Fully 80 percent of known tuberculosis cases in New York City, concentrated among minorities and the homeless, were traced back to prisons (Pablos-Mendez, 2001).

Broader Consequences of a Racially Biased Society

In many cases, differences in racial outcomes are at least partially explainable by differences in the behavior of individuals. In the domain of criminal justice, for example, there is an overrepresentation of nonwhite youth across all stages of the juvenile justice system (National Research Council and Institute of Medicine, 2001). According to self-report data, victimization surveys, and arrest and conviction statistics, black youths show high rates of committing serious offenses compared with white youths. Not surprisingly, these disparities in behavior led to a public discussion focused on individual behavioral choices rather than on past discriminatory processes.

The panel understands that individuals must be held responsible for their actions in the criminal justice system as well as in the education system or the labor market. Individual actions, however, do not occur independently of the larger social and economic context. Certain behaviors by members of disadvantaged racial groups may arise in response to patterns of social and institutional behavior in a racially biased society. Evidence suggests that some behavioral differences may develop over time with differential exposure to risk factors or in reaction to past incidents of discrimination, bias, and exclusion (Cook and Laub, 1998; Sampson and Laub, 1997; Sampson and Lauritsen, 1997; Wilson, 1987). Furthermore, norms and traditions can be affected by incentives (Hobsbawm, 1992).

For instance, frequent and prolonged negative interaction between police and residents in disadvantaged communities can contribute to the overrepresentation of nonwhite youth in the juvenile justice system (Fagan, 2002; National Research Council and Institute of Medicine, 2001). Bachman (1996) found that police respond more rapidly to robberies and aggravated assaults committed by a black offender against a white victim than to those same crimes committed against a black victim or by a white offender. Bachman also found that police devote greater resources to gathering evidence for black offender–white victim crimes, a finding that suggests blacks are more likely to be arrested and subsequently convicted than whites (National Research Council and Institute of Medicine, 2001). Hence, disparities in behavior may be due in part to historical discrimination and current racial stratification.

Exposure to certain risk factors may also explain racial disparities in behavior. Prolonged exposure to risk and negative social interactions over time can influence life choices and limit future opportunities for disadvantaged racial groups. Nonwhite youths, particularly blacks, are disproportionately subject to risk factors associated with crime, such as poverty, poor health care, parental unemployment, and segregation. Youth who believe they have fewer life opportunities or who feel more alienated from mainstream economic and social institutions are probably more likely to engage in risky and self-destructive behaviors. A society that perpetuates strong racial differentials may communicate to nonwhite youth that they are not likely to succeed within mainstream society, leading them to choose alternative lifestyles.

Social isolation and concentration of poverty can marginalize poor individuals from mainstream society (Wilson, 1987). Such conditions disproportionately affect poor minorities, who, cut off from society, lack access to jobs, to higher education, and to positive role models. Without such access, concentrated poverty becomes more acute, leading to a "concentration effect" in which the most disadvantaged members of society (in this case the poorest minorities) are concentrated disproportionately in the most isolated neighborhoods. Wilson argues that social isolation leads to patterns of behavior "not conducive to good work histories," as high unemployment and dissatisfaction with the limited work available lead to altered norms of behavior, such as involvement with drugs or violence (1987:60).

Substantial research has shown that risky and maladaptive behaviors are strongly promoted in neighborhoods of concentrated poverty, many of which are themselves the products of continued racial segregation (Brooks-Gunn et al., 1997; Massey and Denton, 1993). Neighborhoods of concentrated disadvantage, in which a disproportionate share of minorities are disadvantaged or regularly treated with official suspicion, may foster cynicism toward authority and promote illegal deviant behavior (Sampson and Lauritsen, 1997). Furthermore, compounded effects may lead to large differences in future outcomes. For instance, small racial disparities at almost every stage in the juvenile justice process may be compounded through the system (National Research Council and Institute of Medicine, 2001). Thus, the outcome that blacks are disproportionately overrepresented among youth sentenced to correctional institutions—the final stage of the process—may partly result from differential treatment at earlier stages.

Current measures of discrimination that focus on identifying whether discrimination is occurring in a particular domain at a given point in time cannot capture such feedback effects, by which past discrimination affects attitudes, expectations, and behaviors, leading to ongoing and ever widening disparities in outcomes over time. It may be very difficult in such situations to identify empirically a "primary cause" or to measure the share of a differential outcome that is due specifically to past racial discrimination. Yet even if measurement is difficult, it is clear that some adverse outcomes for nonwhites, even when based on freely made personal choices, may partially reflect current and past discrimination that should concern society and motivate the need for research and measurement.

MODELS AND THEORIES OF CUMULATIVE DISADVANTAGE

In most cases, researchers take the results from previous generations or from earlier in a person's lifetime as given and model current behaviors conditional on the past. More dynamic models—particularly those in which past discrimination in some way makes current discrimination more likely—are relatively rare. Here we briefly discuss three theoretical approaches used within three different fields of study that focus on questions of cumulative disadvantage and discrimination: (1) life-course models (criminal justice), (2) ecosocial theory (public health), and (3) feedback models (labor market). It will quickly be apparent that these three approaches (each developed largely independently within separate literatures) have certain elements in common. We present these models not because we think they provide completely satisfactory ways to model the dynamic nature of cumulative discrimination but because they provide possible starting points for future research.

Criminal Justice: A Life-Course Theory of Cumulative Disadvantage

Life-course theory posits that social and historical contexts influence and shape experiences throughout a person's lifetime. Elder (1974, 1975, 1985, 1991, 1998) has done extensive research on the societal influences that shape people's lives from childhood through adolescence and finally adulthood. One challenge of using this perspective is in separating out the effects of the social and historical contexts when examining how current behaviors affect future outcomes in a person's life.

In the criminal justice domain, Sampson and Laub (1997) propose a life-course theory of cumulative disadvantage, which posits that behavior (e.g., criminal delinquency) can affect certain social outcomes (e.g., failure in school or poor job stability) and influence future behavior (e.g., adult criminal activity). Juvenile delinquency, for example, is often linked to adult criminal behavior, as well as other deviant behaviors, such as excessive drinking, traffic violations, and domestic conflict or violence. The developmental framework of Sampson and Laub (1997:135) for understanding continued criminal behavior is based not only on individual behavior but also on "a dynamic conceptualization of social control over the life course." They believe cumulative disadvantage is the result of negative interactions with various key institutions of social control—family, friends, school, and the criminal justice system—that can exacerbate delinquent behavior.

Sampson and Laub argue that cumulative disadvantage results in negative consequences and social sanctions that limit life chances. Thus, societal reactions to criminal delinquency may lead to further deviance, creating a snowball effect: Early delinquency can have negative consequences—arrest, conviction, and incarceration—that limit later opportunities and affect future life chances. Early criminal conviction and incarceration may disrupt schooling and often lead to poorer employment prospects and job instability later in life (Bondeson, 1989; Freeman, 1991; Hagan, 1993; Kasarda and Ting, 1996). Moreover, the length of juvenile incarceration is predictive of subsequent job stability, even after controlling for prior criminal behavior or other delinquencies, such as excessive drinking (Sampson and Laub, 1993).

This model does not directly address the effects of discrimination, although it is apparent that discrimination in the processes that lead a young person to be labeled "deviant" (in the schools or in the juvenile justice system) can contribute to these negative effects. Sampson and Laub (1997) present a theoretical discussion, without attention to how that theory might be tested empirically. The model is complex, with a host of variables that are difficult to measure. It is not obvious how one would identify and trace the causal factors involved through actual longitudinal data. The model is also quite specific to one particular type of disadvantage—related to the labeling and treatment of adolescent offenders—and is thus not directly applicable to a large area of cumulative disadvantage or discrimination.

Public Health: Ecosocial Theory

As in criminal justice research, there is growing recognition in the domain of epidemiology and public health of the importance of the life-course perspective (see Barker, 1998; Kuh and Ben-Shlomo, 1997). In public health, this approach emphasizes how "health status at any given age, for a given birth cohort, reflects not only contemporary conditions but embodiment of prior living circumstances, in utero onwards" (Krieger, 2001:695). Research on health from the life-course perspective examines cross-generational effects of economic deprivation and discrimination, such as how health deficits among African American mothers in poverty (over their life course) affect the well-being of their infants (see, e.g., Lillie-Blanton et al., 1996; Williams and Collins, 1995). Other research has emphasized that one's own income, which can obviously be dampened by discrimination, has an important influence on one's health (Case et al., 2002; Deaton, 2003).

Krieger (1994) proposes an ecosocial theory of cumulative disadvantage for health status due to discrimination over the life course. This theory is based on the assumption that the disparate social and economic status of dominant and subordinate groups leads to differences in their health status. The ecosocial framework, like life-course theory, examines pathways between social experiences and health outcomes. According to Krieger (1999, 2000), cumulative exposure to discrimination can occur through a variety of pathways, including economic and social deprivation, exposure to toxic substances and hazardous conditions, socially inflicted trauma (such as repeated instances of discrimination), targeted marketing of harmful substances, and inadequate health care. Krieger maintains that these pathways may lead to the embodiment or biological expression of experiences of discrimination. For example, economic deprivation can limit access to affordable and nutritious food, which can lead in turn to later health problems (e.g., high blood pressure). Likewise, residential segregation and inadequate access to quality health care can result in higher infant mortality and morbidity.

A small but growing body of literature examines the somatic and mental health consequences of past exposure to racial discrimination (e.g., Mays et al., 1996; Williams and Williams-Morris, 2000). Williams and Neighbors (2001) discuss some laboratory and epidemiological studies using self-report measures, and Krieger (1999) reviews a range of approaches examining the association between institutional discrimination (e.g., residential segregation) and health outcomes within a population. Because this empirical literature is some of the only research linking past experiences of discrimination in one domain with adverse outcomes in another, we describe it further here; as discussed below, however, it may be difficult to draw causal conclusions from much of this work.

Typical laboratory studies in this area use mental imagery, film portrayals, or real-life perceptions of discrimination to measure the effects of exposure to racial bias on health outcomes (see Williams and Neighbors, 2001, for references). For instance, Blascovich et al. (2001) conducted a laboratory experiment in which they manipulated the saliency of stereotype threat (i.e., the threat of being perceived stereotypically) for black participants. Blacks who faced high (versus low) stereotype threat were more likely than whites to show increases in blood pressure. As discussed in Chapter 6, these types of laboratory studies cannot describe the actual occurrence of discrimination over long periods of time, and the findings obtained are not easily generalized to the broader population. Nonetheless, such studies can provide an indication of the explanatory mechanisms that may link past discrimination to current health problems.

Other researchers use statistical methods to relate past experiences of racial disparity and discrimination to current health outcomes. Krieger (1999) notes that the basic strategy is to adjust for factors, such as socioeconomic status, that may explain the observed disparity, then infer discrimination as a possible explanation for any remaining disparity. Williams and Collins (1995) and Lillie-Blanton et al. (1996) review the evidence from studies examining socioeconomic status and racial disparities in health outcomes (e.g., infant mortality, hypertension, and substance abuse). Using self-reported information on past experiences of discrimination, Krieger (1990), Krieger and Sidney (1996), and others (for a review, see Krieger, 1999; Williams and Neighbors, 2001) have found that exposure to discrimination is positively related to higher levels of chronic high blood pressure and hypertension in blacks. For instance, Krieger and Sidney (1996) used large-scale survey data from the multiyear Coronary Artery Risk Development in Young Adults study to examine the association between self-reported experiences of discrimination and blood pressure.

The problems with such approaches are discussed in Chapters 7 and 8. Studies that relate past racial disparities to current health outcomes may not account for unmeasured factors, such as diet and exercise, that may be correlated with race and the observed outcome but that may not be due to discrimination. Analysis that relies on self-reported past measures of discrimination may also be difficult to interpret in any causal way. People who experience high levels of stress may perceive more discrimination or may misattribute nondiscriminatory behavior to discrimination, overestimating the effect. Krieger (1999) notes a variety of problems with the use of self-reports on past discrimination in the health literature.

This health-based ecosocial perspective on the impact of discrimination has many similarities to the life-course theories of criminal justice outcomes. Both focus on differences in treatment that may have long-term behavioral and outcome implications. The ecosocial literature focuses much more on the impact of cumulative discrimination (as opposed to cumulative disadvantage) and provides a clear theoretical discussion of the pathways by which discrimination per se can affect health outcomes over time.

Krieger (1999), in particular, offers some ways to study exposure to discrimination and its effects on health outcomes. She suggests better measures, including experimental studies, in-depth interviews, and large-scale surveys, for capturing exposure to discrimination as well as cumulative exposure over the life course. She emphasizes that these measures should include the level and context of discrimination as well as the onset, frequency, and length of exposure. Williams et al. (2003) also lay out a research agenda for future work. Several researchers have studied the impact of racial discrimination on health outcomes and have made suggestions for improving approaches to measure discrimination in health care (e.g., Darity, 2003; Harrell et al., 2003; Krieger, 2003; Williams et al., 2003). These researchers are careful to note that much of the work in this area is in its infancy, and additional work is required to identify the best methods to measure these associations.

Labor Market: Feedback Models

Feedback effects—whereby past discriminatory events may change future behavior and increase the likelihood of future discrimination—are one way to examine cumulative effects over time; indeed, behavioral feedbacks are embedded in the life-course and ecosocial theories described above. Because of the difficulty of identifying and measuring feedback, there is little empirical work in this area (for exceptions, see Johnson and Neal, 1998; Weiss and Gronau, 1981). This paucity of research makes it difficult to trace the extent to which aggregate outcome differences may be influenced by past discriminatory incidents.

Within the field of labor economics, many researchers have emphasized the importance of feedback effects in analyzing gender and racial discrimination and have developed models of how such effects may occur (e.g., Arrow, 1973; Blau, 1977; Blau et al., 1998; Johnson and Neal, 1998; Lundberg and Startz, 1983, 2000; Weiss and Gronau, 1981). Blau et al. (1998:214) explain the cycle of feedback effects in the labor market for women: "Discrimination against women in the labor market reinforces traditional gender roles in the family, while adherence to traditional roles by women provides a rationale for labor market discrimination." Even a small amount of discrimination can have large effects if women are discouraged from investing in skills, are more likely to opt out of the labor force, and are more likely to rely on their husbands for economic support, hence reinforcing gender roles at home. Policies that help decrease discrimination will also have a feedback effect "as the equalization of market incentives between men and women induces further changes in women's supply side behavior" (Blau et al., 1998:214).

Weiss and Gronau (1981) examine the interaction of labor force participation and wages at different stages in the life cycle and the implications for earnings differences by sex. They posit that earnings in the labor market depend on past participation and investment patterns as well as future participation plans. They also argue that "differences in earnings growth reflect differences in participation plans" (p. 616). Thus, women who expect to participate less in the labor market over time will invest less in raising their earnings capacity. In part, earnings differentials by sex or race may be explained by differences in human capital; however, discrimination may also play a role. For instance, discrimination against women in the labor force can affect patterns of participation or investment. Moreover, expected discrimination may lead to more labor force exits and longer periods spent outside the labor force.

Others have argued that blacks who anticipate lower future returns to skills—possibly as a result of discrimination—may invest less in acquiring those skills (Arrow, 1973; Coate and Loury, 1993; Lundberg and Startz,

1983). The result may be a self-fulfilling prophecy among blacks that perpetuates prejudice, limits opportunities (Krueger, 2002), and sustains racial disparities in the labor market. For instance, Johnson and Neal (1998) note a racial disparity in the number of hours worked by young black and white employees with similar skills. This disparity has a cumulative effect in that differences in weeks of past work experience contribute to the black–white earnings gap. Differences in past work experience may be the result of limited access to employment or job networks but may also be the result of employer discrimination. Moreover, black disadvantage in access to job networks may itself be the result of employer discrimination and may persist even when discrimination is no longer present. Thus, feedback effects may yield negative consequences for black workers who work less because of the lower rewards to work and who subsequently earn less over time. This result is in line with other findings that individuals who experience discrimination engage in behaviors to avoid potential discrimination in the future (Essed, 1991; Feagin, 1991).

An alternative approach is offered by Lundberg and Startz (2000), who model persistence in racial differentials by allowing feedback between individual skill acquisition and community influences. They refer to their model as a model of human capital externalities. In this framework, impoverished communities have less social capital; this in turn affects the human capital acquired by individual members of the community. The result is the persistence of racial differentials, even in the absence of explicit discrimination.

In contrast to the life-course or ecosocial theories discussed above, these labor market theories are more focused and less sweeping in the phenomenon they purport to describe. They tend to provide a clear description of how a particular type of behavior or incentive at one point in time influences behavior at another point in time. They are more mathematically defined, with feedback effects modeled in precise ways. These properties provide a more satisfying description of the particular phenomenon addressed by a theory, but they can limit generalizability. There have been efforts to estimate and measure these feedback effects within the labor market literature; as in other areas, however, it is challenging to measure the right variables and to resolve the identification issues involved in tracing actual discrimination effects over time.

MEASURING CUMULATIVE DISCRIMINATION

In earlier chapters, we discussed the major difficulties involved in measuring credibly and accurately the impact of discrimination within a domain at any point in time. It is even more difficult to measure cumulative effects. This section does not provide a definitive assessment of how to measure cumulative discrimination; rather, we discuss a variety of possible approaches. As noted above, this discussion should be viewed as a suggested research agenda that might be pursued by those interested in trying to determine the importance of cumulative effects relating to discrimination.

Why Measuring Cumulative Discrimination Is Difficult

To measure the cumulative effects of discrimination, at least three things are required. *First, a model and a theory of how cumulative discrimination might occur are needed.* The theory should account for how a particular discriminatory behavior (or behaviors) will influence future behaviors and outcomes and should trace how cumulative discrimination is transmitted across generations, across domains, or over time within a domain. In the previous section, we described three efforts to construct such models to describe dynamic processes within the criminal justice system, the health care system, and the labor market. Effective models of dynamic and long-term processes are still highly limited, however, and much work remains to be done in this area.

Second, one needs to have the longitudinal data necessary to measure effects over time. Meeting this need is most challenging in cross-generational models, which require very long-term data on families. Yet even looking at sequential events over time within a single domain may require extensive longitudinal data on the interactions and activities of an individual. Such data are expensive and difficult to collect; for example, it is difficult to avoid serious attrition problems in long-term longitudinal data sets. Without longitudinal data, progress on the measurement of cumulative discrimination or disadvantage will not be possible. Hence, maintaining the quality and continuity of existing longitudinal data sets is highly important for this area of research.

For instance, the National Longitudinal Survey of Youth provides long-term information about two cohorts of young men and women—one cohort aged 14 to 22 in 1979 that was followed annually through 1994 and biannually since then and another cohort aged 12 to 16 in 1997 that has just started being interviewed annually. These data provide extensive information on family background, expectations, and psychological well-being, as well as detailed year-by-year information on employment, income sources, and living arrangements. This data set has been used extensively to study dynamic processes that affect young people's behavior over time. Other long-term longitudinal data sets that have been used in similar ways include the High School and Beyond data and the Panel Survey of Income Dynamics. Although all of these long-term data sets have limitations, reinterviewing the same people as they become older is necessary to enable credible analysis of almost any question about the impact of past experiences on future choices and behaviors.

Third, in any cumulative process, one needs to be able to identify credibly when exposure to discrimination is occurring. This is often a significant challenge; addressing it requires either direct information on discriminatory behavior or an exogenous source of variation in the conditions that would affect discriminatory behavior. As discussed in Chapters 5 through 8, identifying when discrimination has occurred is often extremely difficult. In the context of a cumulative model, one needs to identify not only the initial incident of discrimination but also (when multiple such incidents may be occurring over time) future incidents of discrimination. This approach would allow one to separate an initial incident of discrimination that affects future outcomes but does not recur from a sequence of discriminatory incidents (that may be causally related to each other) with effects that build cumulatively.

The remainder of this section lays out four possible approaches to identifying and measuring the cumulative effects of discrimination. In each case, we cite a few studies as examples, but even these are typically not very satisfying examples. With a few exceptions, the studies we cite do not themselves claim to be measuring cumulative discrimination. Hence, this section is much less a review of *how* to measure cumulative discrimination than a set of ideas about how one *might think* about measuring cumulative discrimination.

Tabulating Outcomes Over Time

Cross-sectional or longitudinal data can be used to examine widening differentials over time among different groups. Such studies can provide at least potential evidence on the occurrence of cumulative or feedback effects that sequentially worsen outcomes for a certain population. As discussed in Chapter 8, however, longitudinal data are essential for capturing cumulative effects over time for the same individuals.

In the education domain, for example, Phillips et al. (1998) use cross-sectional and longitudinal data from eight national surveys to examine black–white differentials in academic achievement over various grade levels. Including a dummy variable for race, they observe how the race effect is reduced as other variables and their coefficients are included and trace this effect over time. Black students who start school with academic skills comparable to those of the average white student in first grade learn less than the average white student, resulting in a substantially larger negative race effect by the twelfth grade. For instance, Phillips et al. note that the vocabulary scores of black 6-year-olds match those of white 5-year-olds. However, the vocabulary skills of black 17-year-olds are comparable to those of white 13-year-olds (Jacobson et al., 2001; Phillips et al., 1998). During every year of schooling, black students learn less than their white counterparts.

Phillips et al. (1998) note that views about how to measure and interpret the black–white achievement gap vary. That gap is, however, at least consistent with the possibility of cumulative discriminatory effects within the education system, although it provides no direct evidence of discrimination in the schools per se. For instance, research on "summer fallback" (Entwisle and Alexander, 1992, 1994; Entwisle et al., 1997; Heyns, 1978) suggests that the achievement gap widens during the summer when school is out, not during the school year (see Farkas, 2003, for further discussion). Although this result suggests that in-school effects may not be the primary cause of the black–white achievement gap, schools may still play a role in perpetuating the gap.

Investigating racial gaps in outcomes over time requires good data. Robert Hauser (University of Wisconsin-Madison, personal communication) suggests collecting larger sets of observations using direct tests of discriminatory behavior in well-defined settings. One approach is to regularly conduct experimental audit studies across various domains and to trace effects across domains. For instance, Pager (2002) uses matched pairs to estimate the effects of race (being black versus white) and having a criminal record on the likelihood of obtaining an entry-level job. She finds that having a criminal record yields significantly fewer job opportunities for black compared with white testers. Such entry-level racial differences have cumulative effects over time as a result of differential returns to experience. Calculating experience–wage profiles among different populations in the labor market may reveal something about cumulative disadvantage (if not cumulative discrimination).

Identifying Exposure to Discrimination Over Time

One way of identifying discriminatory incidents over time is to use self-reported data on past incidents of discrimination. Conducting longitudinal studies that include validated self-report measures of discrimination, as well as other key variables (e.g., socioeconomic status), can permit the study of long-term effects of discrimination on such outcomes as health (Krieger, 1999). Krieger and Sidney (1996), for example, use a self-report method to assess experiences of discrimination in multiple situations (e.g., at school, at work, obtaining medical care, obtaining housing) and to examine the association of discrimination with hypertension. Within the labor market, Neumark and McLennan (1995) investigate the effects of reported discrimination on women's labor market participation and outcomes. They find that women who report discrimination are more likely to change employers but find little effect on long-term wage growth.

We have already discussed the limitations to using self-reported data as a measure of discrimination in Chapter 8. Such measures can be ambiguous and difficult to interpret; they can either overestimate or underestimate discrimination. Stating her concerns with these issues, Susan Murphy (University of Michigan, personal communication) suggested one might use multiple measures of exposure to discrimination and link these measures with specific outcomes. She also advised collecting as much information as possible about individual, situational, and contextual reasons for a person's exposure to discrimination (e.g., personal appearance or being female in a male-dominated occupation). This information may help exclude alternative explanations for certain outcomes.

To assess both cumulative and delayed effects of exposure to discrimination, one must adjust for any compositional differences between groups with higher exposures. For example, there may be some situations—such as being a woman in an almost-all-male occupation or being a black man in an almost-all-white-male occupation—that put one at greater risk of experiencing discrimination. Adjustment for selectivity that accounts for other differences between groups that choose different occupations is particularly crucial when exposures occur over time; adjustment for compositional differences is then required repeatedly. It is also important not just to measure the effect of small exposures at each time point: As discussed above, effects at any one time may be small, but the total effects of long-term exposure can be cumulative and more than just the sum of many small exposures (see also Chapter 7).

Although much of the existing (sparse) literature relies on self-reports of discrimination, it is important to develop other methods for assessing when discrimination occurs. Research in social psychology has shown that people may underestimate the frequency of discriminatory events in their own life compared with discrimination against their group (Crosby, 1984; Taylor et al., 1991). At the same time, however, the extent to which people perceive events as discriminatory is likely to have effects on various aspects of their lives regardless of the so-called objective occurrence of such an event. Ideally, then, methodologies should include both self-reports and implicit or observational assessments of discriminatory actions.

One might use group-level experiences of discrimination as a means of assessing individual reports of discrimination. These experiences might include evidence on residential segregation and population-level expressions of empowerment, including representation in government. Low reported levels of individual discrimination in the context of substantial institutional exclusion would suggest problems with individual reports.

Estimating Current Outcomes from Past Events

The most common approach to measuring cumulative effects across domains or over time is to use past events and outcomes as determinants of current outcomes. Such estimates may use cross-sectional or longitudinal data. The previous section addressed the possibility of such analysis when one has actual information on past incidents of discrimination (see Chapter 7). But having information on the presence or absence of discrimination in the past is rare. Typically, one can merely control for past outcomes that create current predetermined variables, such as educational or skill levels, current health status, or past criminal record. In controlling for these past events, one is typically unable to identify how much of any past outcome is due to discrimination and hence how much past discrimination may be affecting current outcomes.

For instance, in estimating the determinants of employment or wages in the labor market, controlling for outcomes within the educational system is standard (Blau and Kahn, 1997). The coefficients on education are interpreted as the return to human capital (skill levels) in the labor market. The causal factors that go into determining that level of skill are taken as given. If discrimination in the educational system is impeding the skill level achieved by racially disadvantaged students, this is taken as a predetermined factor in the labor market. The emphasis of such an equation is not on measuring the cumulative effects of discrimination but on determining whether there is any evidence of discrimination within the labor market only. This is clearly a useful and important question, but it is not the question one might ask when focusing on the effects of cumulative and overtime exposure to discrimination during the life course.

The potential importance of these cross-domain effects is reviewed by Neal and Johnson (1996), who argue that differences in skills before entering the market explain most of the racial gap in wages. Taken at face value, this research suggests that understanding racial differentials in the labor market requires an understanding of the processes that produce pre-labor market skill differences. Goldsmith et al. (2000) argue that Neal and Johnson's results are flawed, and they include measures of motivation as preferred control variables. However, their findings raise the question of where individual motivation is learned and suggest that family and school backgrounds might influence important behavioral characteristics that are fundamental to labor market performance.

It should be possible to estimate the approximate magnitude of more cumulative effects of discrimination through multiple regressions at different stages in a process. For instance, one could use a two-step process, first measuring the effect of discrimination on outcome variable 1 in domain 1 (say, discrimination on educational outcomes) and then estimating the effect of outcome variable 1 on a (future) outcome variable 2 (say, employment outcomes) in domain 2. Assuming a credible measure of the impact of discrimination in the first stage, one could use these two results to impute the effect of discrimination in domain 1 on outcome 2. For example, one could estimate the effect of discrimination on high school completion rates and estimate the impact of high school completion on wages. Next, one could impute the impact of discrimination in education on wages. As discussed in Chapter 7, however, drawing causal conclusions about discrimination by fitting regression models to observational data requires strong assumptions.

Using Identifying Information on the Occurrence of Discrimination

A final approach is to find identifying information that signals discrimination from some earlier time period and that can be directly entered into an outcome estimate at a future time period. For instance, Sacerdote (2002) assesses the impact of slavery on literacy and occupations across generations. Using census data from 1880 and 1920, he examines the effect of slavery on outcome differences for former slaves and free blacks and for their children and grandchildren. This approach reflects a tradition in sociology that dates back to Duncan's (1968) classic paper examining the extent to which the economic and educational disadvantages of the current generation of blacks can be explained by the economic and educational disadvantages of their parents. This kind of model is a special case of Duncan's "status attainment" or "life-cycle" model of attainment; it is used in various areas to examine cumulative effects (e.g., Phillips et al., 1998). However, Sacerdote (2002) notes a lack of research on intergenerational effects because few longitudinal data sets provide information on three or more generations of family members.

In another example, Card and Krueger (1992) examine the effect of school resources on wages, using state school desegregation dates as an instrument for improvement in schools among black children in southern states. Differences in the resources available to black versus white schools in a community can be taken as a measure of discrimination (although this is not the interpretation or focus of their paper). Past generations that lived under the old segregated schools may have experienced more discrimination, and the impact of school desegregation can be used as a measure of the impact of reduced discrimination on educational outcomes.

Finding a credible variable for a policy or a past experience that was clearly discriminatory can be challenging, although policy changes over the past several generations might signal a reduction in discrimination from one point in time to another. Although a number of papers look at the immediate impact of policy changes (such as the adoption of Title VII of the

1964 Civil Rights Act), it may also be possible to examine over-time and cumulative effects of discriminatory policies by comparing changes across generations that lived before and after these policy changes.

SUMMARY, CONCLUSION, AND RECOMMENDATION

The discussion in previous chapters focused on single instances of racial discrimination at a specific point in time within a particular domain. In this chapter, we explore the possibility of cumulative effects of discrimination—occurring over time and across domains—that might be missed using standard measurement approaches. Estimating the cumulative effects of discrimination over time is a difficult and challenging task and only a limited number of studies attempt to do so. Some theories of discrimination and disadvantage describe ways in which individual behaviors, societal influences, institutional practices, and exposure to risk may cumulate over time to affect future life choices and opportunities. However, both the theoretical and the empirical work in this area is in its infancy.

We suspect that the cumulative effects of discrimination, although seriously understudied, may be important. Of course, to prove or disprove the importance of cumulative effects, there is a need for research that credibly measures the presence or absence of such effects. To investigate cumulative impacts of discrimination more effectively, progress is necessary in several areas. First, there is a need for better theoretical work on how to conceptualize the dynamic and time-dependent effects of cumulative discrimination. Second, there is a need for better longitudinal data on different outcomes and events in a variety of domains, perhaps even across generations. Third, there is a need for creative ways to identify and estimate cumulative effects of discrimination over time and across domains (e.g., self-report and multiple measures).

It is possible that much of the current evidence on discrimination—even when credibly estimated—may be of limited value in answering the question "What is the net effect of discrimination in American society?" Discrimination may occur at one stage in a process (e.g., labor market) and contribute only a small amount to racial differences in immediate outcomes. At later stages, however, the initial discrimination may have effects that cumulate over time, but current measures may not capture those effects. Because of the possible dynamic processes that may lead to cumulative disadvantage, it is difficult to determine the extent to which observed aggregate differences by race are due to discrimination. Particularly if discrimination at one point in the life course is magnified over time, whether because of individual behavioral responses or because of institutional practices, many current measures of discrimination are insufficient to identify the overall impact of discrimination on individuals.

As we have noted throughout this chapter, a key element in any research on cumulative discrimination is the availability of good longitudinal data. Therefore, studying the cumulative effects of discrimination requires the collection of longitudinal data that provide repeated measures for the same individual over time. Although existing longitudinal data sets are necessarily limited in the data they provide to investigate discrimination (or any other topic), they contain long-term information about behaviors and outcomes over time and across generations that allows the estimation of more dynamic models.

Conclusion: *Measures of discrimination from one point in time and in one domain may be insufficient to identify the overall impact of discrimination on individuals. Further research is needed to model and analyze longitudinal and other data and to study how effects of discrimination may accumulate across domains and over time in ways that perpetuate racial inequality.*

Recommendation 11.1. Major longitudinal surveys, such as the Panel Study of Income Dynamics, the National Longitudinal Survey of Youth, and others, merit support as data sources for studies of cumulative disadvantage across time, domains, generations, and population groups. Furthermore, consideration should be given to incorporating into these surveys additional variables or special topical modules that might enhance the utility of the data for studying the long-term effects of past discrimination. Consideration should also be given to including questions in new longitudinal surveys that would help researchers identify experiences of discrimination and their effects.

12

Research: Next Steps

Our report discusses the challenges of measuring racial discrimination in a range of social and economic domains. Establishing that overt or subtle forms of discrimination have occurred and the consequent effects on outcomes requires careful and thorough analysis to rule out or limit alternative explanatory factors. In much research to date, the data and analytical methods make it difficult to justify the assumptions of the underlying theoretical model. Moreover, many statistical and survey-based analyses never articulate an explicit model, which makes it difficult to judge the adequacy of the data and analysis to support the study findings. Laboratory experiments, while often better justified, cannot in and of themselves measure the contribution of discrimination to differential outcomes in a real-world setting.

Although it is difficult to measure racial discrimination, it is possible to conduct important, appropriate research in this area that adds to our knowledge. Some laboratory and field experiments, statistical analyses of observational data, evaluations of natural experiments, and survey measures of discriminatory attitudes and reported experiences of discrimination have produced useful results pertaining to particular types of possible discrimination within a domain or process. To make further progress, we believe it will be necessary for funding and program agencies to support studies that cut across disciplinary boundaries, make use of multiple methods and types of data, and analyze racial discrimination as a dynamic process rather than as a point-in-time event. It will also be necessary for program and research agencies to identify priority areas for which research on the possible role of racial discrimination is most needed and to further the development of useful data sources for measurement purposes.

Our efforts as a panel concentrated on an in-depth exploration of concepts and methodological approaches to measuring racial discrimination. Within the scope of our charge and resources, we could not take the next step of developing a detailed agenda in any domain for further research to inform policy making and public understanding. What we undertake in this short concluding chapter is to suggest ways in which program and research agencies might build a research agenda that is directed to priority needs for measuring racial discrimination.

PROGRAM AGENCIES

Program agencies that are charged to monitor and investigate discrimination complaints, such as the U.S. Department of Education Office of Civil Rights, the U.S. Equal Employment Opportunity Commission, and others, have a direct interest in the measurement and understanding of racial discrimination. These agencies could benefit most directly from improved data and research in relevant domains of interest to them. Other agencies that design and operate programs that may be directly affected by the presence of discrimination and by antidiscrimination laws and regulations should also have an interest in discrimination research (such agencies exist in the U.S. Departments of Health and Human Services, Housing and Urban Development, Justice, Labor, and others).

Priority Research Topics

An initial step in furthering useful research for program needs is for agencies to identify the subset of outcomes and processes in which racial discrimination may occur that are of most importance from the agency's policy perspective. This is not a trivial task. It is crucial, however, to framing a cost-beneficial research agenda, given the substantial time and effort that would likely be required to obtain appropriate data and conduct useful analyses on even a single topic.

Because resource limitations will necessarily constrain research and data collection, program agencies should subject their list of priority research areas to careful evaluation regarding feasibility and costs. We strongly urge that agencies not limit their determination of feasible priority projects to a particular disciplinary perspective or type of analytical method or data. Narrowing a methodological focus too early could well lead to conclusions that do not stand up when subjected to other kinds of analyses.

As a hypothetical example, consider racial discrimination in the employment domain, which clearly presents many questions of policy and public interest. A review of the labor market literature, as well as an analysis of program agency data on discrimination complaints, could help identify priority topics. For instance, some research has indicated (see Chapter 11) that equally qualified nonwhite and white college graduates are hired at similar starting salaries but that nonwhites become increasingly disadvantaged with regard to earnings over time. These results suggest that research on employer decision processes related to job training and promotion could merit greater attention in the near future than, say, replication of studies of factors in initial hiring decisions.

The next step is to develop a detailed research plan. In Chapter 7, we argued that statistical information on racial gaps in outcomes will rarely be adequate to support conclusions about the role of racial discrimination in the absence of a detailed understanding of the decision processes of decision makers, including information on what knowledge is available to them and what knowledge they bring to bear in making particular types of decisions. In the labor market example, this would mean understanding the processes by which hiring or promotion occurs and the information available to employers in making employment or promotion decisions. Focused case studies of employer decision processes may be needed to provide the requisite depth of understanding of employer behavior to permit subsequent statistical analysis.[1]

To be most useful and cost-effective, focused studies of decision-making processes should be informed by theoretical models of the ways in which discrimination might occur. Especially because discrimination may take subtle, as well as overt, forms, it is important to have a theoretical framework to guide the data to be collected in case studies. In the labor market example, such a framework could help determine which actors in a firm to interview; what kinds of institutional practices, policies, and procedures to learn about; and what other information to collect.

In developing a theoretical framework, researchers could usefully review the existing literature of laboratory experiments about discriminatory attitudes and behaviors and the kinds of situations in which attitudes are most likely to lead to race-based discriminatory treatment. For instance, an economics approach to studying discrimination could be enhanced by psychological insights derived from empirical results of laboratory experiments, as well as from psychological concepts about the functioning and sources of discriminatory attitudes and behavior. If laboratory results are not suffi-

[1]The same arguments apply to analysis of the contribution of race-based discrimination to outcomes in other domains; namely, the likely need for focused case studies of relevant decision processes (e.g., admissions to colleges and universities; applications for loans to banks and government agencies; or access to health care at hospitals, clinics, doctors' offices, and other venues) to inform data collection and the construction of sound statistical models.

ciently focused on the decision making relevant to an agency, then additional experiments could be commissioned to fill in gaps. Such experimentation would require that laboratory researchers develop methods for obtaining participation from people in the work world and other venues outside of academia.

Consideration of appropriate concepts and review of pertinent laboratory results should help suggest the types of data that are most needed for informative analyses of observational data with statistical models. For example, case studies might justify adding questions to cross-sectional and longitudinal surveys on self-reports of discrimination, or they might suggest collecting information on specific characteristics related to the decision process, such as (again, a labor market example) whether the employee was recommended by another employee for an open position, what kinds of testing and interviews were required, and so on.

With data in hand, and with well-developed models of decision processes, agencies would be poised to create a research agenda around questions they would like to present to researchers well versed in statistical analysis methods that are appropriate for assessing evidence of discrimination and its impacts. Program agencies should also consider the possibilities of field studies that bring scientific evaluation techniques to real-world decision-making examples. The use of audit studies within the U.S. Department of Housing and Urban Development, for example, has helped support the claim that ongoing housing market discrimination occurs in the housing search process, which suggests the importance of ongoing enforcement of open-housing policies. Agencies in other domains should consider the possibility of field or audit studies in their own areas of interest.

The work we outline above would require collaboration of scholars from multiple disciplines, including economists, sociologists, social psychologists, and survey researchers. In some situations, ethnographers and cultural anthropologists could also contribute much-needed expertise for designing and conducting the case studies of employers or other decision makers to obtain the richest data possible.

Facilitating Data Access and Use

Another way in which program agencies could facilitate a cost-effective agenda for research on the possible role of racial discrimination in domains of interest concerns the provision of data. Agencies should first analyze the research potential of their own administrative records, identifying low-cost changes to record requirements that would facilitate analytical use of the data. Concurrently, agencies could work to develop arrangements for reasonably ready access to the data by qualified researchers.

In practice, the development of suitable administrative records data for analysis of racial discrimination, particularly if the research goal is to compare administrative records with survey reports of discrimination events, is likely to present difficult problems. There could well be problems in obtaining access to administrative records, understanding agency reporting systems, and protecting the confidentiality of the data. Nonetheless, because administrative data are maintained for recordkeeping purposes by enforcement agencies and thereby provide a low-cost resource for research, it seems worthwhile to conduct feasibility studies to determine their potential for analytical use, alone and linked to survey data. Such use could not only add to knowledge but also help agencies design more informative records systems for their own enforcement and education programs.

Program agencies could also provide input to the federal statistical system regarding data items that would be useful to include in ongoing household cross-sectional and longitudinal data systems run by statistical agencies. Major longitudinal surveys of cohorts of individuals exist in the domains of labor market experience, education, and health. Such surveys are prime candidates to review to identify cost-effective additions or modifications of questions that would support research on discrimination. Statistical agencies can contribute to the provision of useful data for analysis of discrimination by sponsoring research, as we recommended in Chapter 10, on best practices for obtaining data on racial and ethnic classifications.

Finally, program agencies can play a valuable role in facilitating research evaluation of natural experiments consequent to policy and regulatory changes, by modifying or augmenting administrative records systems, as appropriate. With suitable data, natural experiment evaluations can compare differences in outcomes over time and between individuals affected and not affected by these changes, in ways that can illuminate the possible role of racial discrimination.

RESEARCH AGENCIES

We suggest that research funding agencies, such as the National Science Foundation, the National Institutes of Health, and private foundations, can best leverage their resources by addressing areas of research on racial discrimination that are less apt to be considered by program agencies. They also have a comparative advantage in supporting more basic research and data infrastructure, including support for rich longitudinal data collections.

Within-Domain and Across-Domain Cumulative Effects Studies

Research funding agencies are better positioned than program agencies to support innovative, cross-disciplinary, multimethod research on cumulative disadvantage and the roles that current and past discrimination—

whether in a particular stage of a process or in other domains—may play in causing a set of differential outcomes. They are also better positioned to support innovative studies of the possible role of discrimination in cumulative disadvantage over a lifetime and across generations. Our discussion in Chapter 11 of cumulative disadvantage described the need for and the difficult nature of such studies, of which there are very few examples to date.

To move cumulative effects research forward, it could be useful to build on studies of the possible role of discrimination in differential outcomes in one stage of a process to develop insights about the role of past and current discrimination for a subsequent stage of the same process or for another domain. As one example, field experiments in housing and labor markets might provide a basis for work on subsequent outcomes in those domains, by identifying geographic areas or types of firms for which experimental results suggest particularly strong effects of discrimination at an entry level (seeking an apartment or home, seeking a job). These areas or firms could possibly be further studied to consider the cumulative effects of the initial-stage discrimination on outcomes at subsequent stages (e.g., ability to obtain home equity loans or refinancing, access to training and promotion opportunities).

Research funding agencies could also consider supporting studies of the effects of discrimination in one domain, such as housing, on processes in another domain, such as access to schools. They could consider supporting studies of longer-term discrimination over lifetimes and generations. Such cross-process, cross-domain, cross-generation types of research will necessarily require bringing together researchers from multiple disciplines and perspectives and using various data sets and methods—for example, using laboratory experiments to develop theoretical constructs for paths and mechanisms by which cumulative disadvantage could occur; using case studies and ethnographic research to obtain very rich data on perceptions and experiences of discrimination in a particular population group or community; and using rich panel data to follow population cohorts over time.

There are examples of rich, multidisciplinary, cross-domain research in other areas of inquiry that discrimination researchers might look to for guidance. In particular, a number of multifaceted studies have been conducted in recent years of changes in the well-being of low-income populations following major changes in welfare policies (see National Research Council, 2001b). These studies have combined national surveys, surveys of specific cities and neighborhoods, and in-depth ethnographic research to understand the factors that contributed to a range of social and economic outcomes for low-income families in a period of rapid policy and economic change. Survey data have come from repeated cross-sectional interviews and longitudinal panels and from interviews of welfare case workers in addition to welfare recipients, people leaving the welfare rolls, and other low-income families. Some studies have included field experiments of alternative welfare policies; other studies have taken advantage of natural experiments provided by major changes in national welfare policy and variations in implementation by states.

Panel Data

We have stressed in several chapters the need for rich longitudinal data sets that follow individuals over time and hence permit studies of cumulative disadvantage, as well as studies that delineate paths by which disadvantage—and possible discrimination—occurs. Statistical agencies fund some of the major panel surveys, such as the National Longitudinal Surveys of Labor Market Behavior of the Bureau of Labor Statistics, but many panel surveys are funded by public and private research agencies. These surveys represent significant components of the data infrastructure for social science research. Public and private research agencies interested in facilitating studies of racial discrimination, particularly over long periods of time, can usefully consider ways to augment ongoing and new panel surveys to provide relevant data.

CONCLUSION

Our report has documented the strengths and weaknesses that various methodologies and data sources can bring to the table for measuring racial discrimination. The difficulties of analysis in this area make it daunting for program and research funding agencies to develop focused, cost-effective agendas for research and data collection. We have suggested some strategies for developing future research plans. We urge that research on racial discrimination, whether focused on program agency priorities for analysis of a particular domain or more basic research on cumulative disadvantage, bring multiple perspectives to bear and use multiple methods and data sources. Although current and even past racial discrimination may be only part of the explanation for persistent racial gaps in important domains of social and economic life, it is important for public policy and public understanding to carry out research on the role of discrimination among all of the factors that shape American society today.

References

Ajzen, I.

1991 The theory of planned behavior. *Organizational Behavior and Human Decision Processes* 50:179–211.

2001 Nature and operation of attitudes. *Annual Review of Psychology* 52:27–58.

Alba, R.

1990 *Ethnic Identity*. New Haven, CT: Yale University Press.

1992 Ethnicity. In *Encyclopedia of Sociology*, E. Borgatta and M. Borgatta, eds. New York: MacMillan.

Alba, R., A. Lutz, and E. Vesselinov

2001 How enduring were the inequalities among European immigrant groups in the United States? *Demography* 38:349–356.

Alexander, K.L., D.R. Entwisle, and R. Herman

1999 In the eye of the beholder: Parents' and teachers' ratings of children's behavioral style. In *Contemporary Perspectives on Family Research*, Vol. 1, C. L. Shehan, ed. Greenwich, CT: JAI Press.

Allen, J.

1995 A possible remedy for unthinking discrimination. *Brooklyn Law Review* 61(Winter):1299.

Allen, M.

2003 Bush issues ban on racial profiling. *The Washington Post*, June 18, p. A14.

Allport, G.

1954 *The Nature of Prejudice*. Reading, MA: Addison-Wesley.

Altemeyer, B.

1988 *Enemies of Freedom: Understanding Right-Wing Authoritarianism*. San Francisco, CA: Jossey-Bass.

1996 *Authoritarian Spector*. Cambridge, MA: Harvard University Press.

Altonji, J.G., and R. Blank

1999 Race and gender in the labor market. *Handbook of Labor Economics* 3:3143–3259.

Altonji, J.G., and U. Doraszelski

2002 The Role of Permanent Income and Demographics in Black/White Differences in

Wealth. Yale University Economic Growth Center Discussion Paper No. 850. Available: http://www.library.yale.edu/socsci/egcdp850.txt [January 29, 2004].

Altonji, J.G., and C.R. Pierret
2001 Employer learning and statistical discrimination. *Quarterly Journal of Economics* (February):1–37.

American Civil Liberties Union
2000 Plaintiff's Fifth Monitoring Report: Pedestrian and Car Stop Audit. Available: http://www.aclupa.org/legal/racial_profiling.pdf [January 29, 2004].

Anderson, B.R.
1983 *Imagined Communities: Reflections on the Origin and Spread of Nationalism.* London: Verso.

Anderson, M.
1988 *The American Census: A Social History.* New Haven, CT: Yale University Press.
2000 *Encyclopedia of the U.S. Census.* Washington, DC: Congressional Quarterly Press.

Anderson, M., and S. Fienberg
1999a To sample or not to sample? The 2000 census controversy. *Journal of Interdisciplinary History* 30(1):1–36.
1999b *Who Counts? The Politics of Census-Taking in Contemporary America.* New York: Russell Sage Foundation.
2000 Race and ethnicity and the controversy over the U.S. Census. *Current Sociology* 48(3):87–110.

Anderson, N.B.
1989 Racial differences in stress-induced cardiovascular reactivity and hypertension: Current status and substantive issues. *Psychological Bulletin* 105:89–105.

Antecol, H., and P. Kuhn
2000 Gender as an impediment to labor market success: Why do young women report greater harm? *Journal of Labor Economics* 18(4):702–728.

Anti-Defamation League
2001 *Audit of Anti-Semitic Incidents, 2000.* New York: Anti-Defamation League.

Appiah, K.A.
1992 *My Father's House: Africa in the Philosophy of Culture.* Oxford, England: Oxford University Press.

Arrow, K.
1973 The theory of discrimination. In *Discrimination in Labor Markets*, O. Ashenfelter and A. Rees, eds. Princeton, NJ: Princeton University Press.

Ayres, I., and P. Siegelman
1995 Race and gender discrimination in bargaining. *American Economic Review* 85:304–321.

Bachman, R.
1996 Victim's perceptions of initial police responses to robbery and aggravated assault: Does race matter? *Journal of Quantitative Criminology* 12(4):363–390.

Banton, M.P.
1983 *Racial and Ethnic Competition.* New York: Cambridge University Press.

Bargh, J.A., and T.L. Chartrand
1999 The unbearable automaticity of being. *American Psychologist* 54:462–479.

Bargh, J.A., M. Chen, and L. Burrows
1996 Automaticity of social behavior: Direct effects of trait construct and stereotype priming on action. *Journal of Personality and Social Psychology* 71:230–244.

Barker, D.J.P.
1998 *Mothers, Babies, and Health in Later Life*, 2nd ed. Edinburgh, Scotland: Churchill Livingston.

Barsky, R., J. Bound, K.K. Charles, and J.P. Lupton
2002 Accounting for the black-white wealth gap: A nonparametric approach. *Journal of the American Statistical Association* 97(459):663–673.
Bates, N.A., M. De La Puente, T.J. DeMaio, and E.A. Martin
1994 Research on race and ethnicity: Results from questionnaire design tests (disc. P160-6). In *Proceedings of the Bureau of the Census Annual Research Conference*. Washington, DC: U.S. Bureau of the Census.
Beck, E.M., and S.E. Tolnay
1990 The killing fields of the deep South: The market for cotton and the lynching of blacks, 1882–1930. *American Sociological Review* 55(August):526–539.
Becker, G.S.
1957 *The Economics of Discrimination*. Chicago: University of Chicago Press.
1971 *The Economics of Discrimination*, 2nd ed. Chicago: University of Chicago Press.
Bendick, M., Jr., C.W. Jackson, and V.A. Reinoso
1994 Measuring employment discrimination through controlled experiments. *Review of Black Political Economy* 23:25–48.
Benson, P.L., S.A. Karabenick, and R.M. Lerner
1976 Pretty pleases: The effects of physical attractiveness, race, and sex on receiving help. *Journal of Experimental Social Psychology* 12:409–415.
Bertrand, M., and S. Mullainathan
2002 Are Emily and Brendan More Employable than Lakisha and Jamal? A Field Experiment on Labor Market Discrimination. Graduate School of Business Working Paper, University of Chicago.
Bickel, P.J., and J.W. O'Connell
1975 Is there a sex bias in graduate admissions? *Science* 187:398–404.
Biernat, M., and D. Kobrynowicz
1997 Gender- and race-based standards of competence: Lower minimum standards but higher ability standards for devalued groups. *Journal of Personality and Social Psychology* 72(3):544–557.
Black, D.A.
1995 Discrimination in an equilibrium search model. *Journal of Labor Economics* 13(2):309–334.
Black, D., A. Haviland, S. Sanders, and L. Taylor
2002 Why Do Minority Men Earn Less? A Study of Wage Differentials Among the Highly Educated. Unpublished paper, Carnegie Mellon University, Pittsburgh, PA.
Black, S.E., and P.E. Strahan
2001 The division of spoils: Rent-sharing and discrimination in a regulated industry. *American Economic Review* 91(4):814–831.
Blackwell, C., and J. Mehaffey
1983 American Indians, trust and recognition. In *Nonrecognized American Indian Tribes: An Historical and Legal Perspective*, F. Porter III, ed. (Occasional Papers Series, No. 7.) Chicago: The Newberry Library.
Blair, I.V., C.M. Judd, M.S. Sadler, and C. Jenkins
2002 The role of Afrocentric features in person perception: Judging by features and categories. *Journal of Personality and Social Psychology* 83(1):5–25.
Blank, R.M.
2001 An overview of trends in social and economic well-being, by race. In *America Becoming: Racial Trends and Their Consequences*, Vol. I, N.J. Smelser, W.J. Wilson, and F. Mitchell, eds. Commission on Behavioral and Social Sciences and Education, National Research Council. Washington, DC: National Academy Press.

Blascovich, J., S.J. Spencer, D. Quinn, and C. Steele

2001 African Americans and high blood pressure: The role of stereotype threat. *Psychological Science* 12(3):225–229.

Blau, F.D.

1977 *Equal Pay in the Office*. Lexington, MA: Lexington Books.

Blau, F., and A. Beller

1992 Black-white earnings over the 1970s and 1980s: Gender differences in trends. *Review of Economics and Statistics* 74(2):276–286.

Blau, F.D., and L.M. Kahn

1997 Swimming upstream: Trends in the gender wage differential in the 1980s. *Journal of Labor Economics* 15(1):1–42.

Blau, F.D., M.A. Ferber, and A.E. Winkler

1998 *The Economics of Women, Men, and Work*. Upper Saddle River, NJ: Prentice-Hall.

Blaut, J.M.

1993 *The Colonizer's Model of the World: Geographic Diffusionism and Eurocentric History*. New York: Guillford.

Blinder, A.

1973 Wage discrimination: Reduced form and structural variables. *Journal of Human Resources* 8:436–455.

Blumstein, A.

1982 On the racial disproportionality of United States' prison populations. *Journal of Criminal Law and Criminology* 73:1259–1281.

1993 Racial disproportionality of U.S. prison populations revisited. *University of Colorado Law Review* 64:743–760.

Bobo, L.D.

1983 Whites' opposition to busing: Symbolic racism or realistic group conflict? *Journal of Personality and Social Psychology* 45:1196–1210.

1997 The color line, the dilemma, and the dream: Race relations in America at the close of the 20th century. In *Civil Rights and Social Wrongs: Black-White Relations Since World War II*, J. Higham, ed. University Park: Pennsylvania State University Press.

2001 Racial attitudes and relations at the close of the twentieth century. In *America Becoming: Racial Trends and Their Consequences*, Vol. I, N.J. Smelser, W.J. Wilson, and F. Mitchell, eds. Commission on Behavioral and Social Sciences and Education, National Research Council. Washington, DC: National Academy Press.

Bobo, L.D., and J.R. Kluegel

1997 Status, ideology, and dimensions of whites' racial beliefs and attitudes: Progress and stagnation. In *Racial Attitudes in the 1990s: Continuity and Change*, S.A. Tuch and J.R. Martin, eds. Westport, CT: Praeger Press.

Bobo, L.D., and M.P. Massagli

2001 Stereotyping and urban inequality. In *Urban Inequality: Evidence from Four Cities*, A. O'Conner, C. Tilly, and L.D. Bobo, eds. New York: Russell Sage Foundation.

Bobo, L.D., and R. Smith

1998 From Jim Crow racism to laissez faire racism: The transformation of racial attitudes. In *Beyond Pluralism: Essays on the Conception of Groups and Group Identities in America*, W. Katkin, N. Landsman, and A. Tyree, eds. Urbana: University of Illinois Press.

Bobo, L.D., and S.A. Suh

2000 Surveying racial discrimination: Analyses from a multiethnic labor market. In *Prismatic Metropolis: Inequality in Los Angeles*, L.D. Bobo, M.L. Oliver, J.H. Johnson, Jr., and A. Valenzuela, eds. New York: Russell Sage Foundation.

Bobo, L.D., J. Kluegel, and R. Smith
1997 Laissez-faire racism: The crystallization of a kinder, gentler, anti-black ideology. In *Racial Attitudes in the 1990s: Continuity and Change*, S. Tuch and J. Martin, eds. Westport, CT: Praeger.
Bondeson, U.V.
1989 *Prisoners in Prison Societies.* New Brunswick, NJ: Transaction Publishers.
Bonilla-Silva, E., and T.A. Forman
2000 "I am not a racist but...": Mapping white college students' racial ideology in the USA. *Discourse and Society* 11(1):50–85.
Boozer, M., A. Krueger, S. and Wolkon
1992 Race and school quality since Brown v. Board of Education. *Papers on Economic Activity–Microeconomics*, 269–338.
Borjas, G.J.
1994 Long-run convergence of ethnic skill differentials: The children and grandchildren of the great migration. *Industrial and Labor Relations Review* 47:553–573.
Borjas, G.J., and S.G. Bronars
1989 Consumer discrimination and self-employment. *Journal of Political Economy* 97(3):581–606.
Bowlus, A.J., and Z. Eckstein
2002 Discrimination and skill differences in an equilibrium search model. *International Economic Review* 43(4):1309–1345.
Bowman, P.J., and C. Howard
1985 Race-related socialization, motivation, and academic achievement: A study of black youth in three-generation families. *Journal of the American Academy of Child Psychiatry* 24:134–141.
Boykin, A.W., and F. Toms
1985 Black child socialization: A conceptual framework. In *Black Children: Social, Educational, and Parental Environments*, H. McAdoo and J. McAdoo, eds. Thousand Oaks, CA: Sage.
Braddock, J.H., and J.M. McPartland
1987 How minorities continue to be excluded from equal employment opportunities: Research on labor market and institutional barriers. *Journal of Social Issues* 43:5–39.
Brewer, M.B., and R. Brown
1998 Intergroup relations. In *The Handbook of Social Psychology*, 4th Edition, D. Gilbert, S.T. Fiske, and G. Lindzy, eds. New York: McGraw-Hill.
Brookings Institution
2000 *Government's Greatest Achievements of the Past Half Century.* Washington, DC: Brookings Institution Press.
Brooks-Gunn, J., G.J. Duncan, and J.L. Aber
1997 *Neighborhood Poverty: Context and Consequences for Children.* New York: Russell Sage Foundation.
Brown, T.N
2001 Measuring self-perceived racial and ethnic discrimination in social survey. *Sociological Spectrum* 21:377–392.
Brundage, W.F.
1993 *Lynching in the New South: Georgia and Virginia, 1880–1930.* Urbana: University of Illinois Press.
Bureau of Labor Statistics
2003 *Employment and Earnings*, Vol. 50(1).Washington, DC: U.S. Department of Labor.

Camarillo, A.M., and F. Bonilla
2001 Hispanics in a multicultural society: A new American dilemma? In *America Becoming: Racial Trends and Their Consequences*, Vol. I, N.J. Smelser, W.J. Wilson, and F. Mitchell, eds. Commission on Behavioral and Social Sciences and Education, National Research Council. Washington, DC: National Academy Press.

Campbell, D.T., and J.C. Stanley
1963 *Experimental and Quasi-Experimental Designs for Research*. Chicago: Rand McNally College Publishing Company.

Card, D., and A. Krueger
1992 School quality and black-white relative earnings: A direct assessment. *Quarterly Journal of Economics* 107:151–200.
1994 Minimum wages and employment: A case study of the fast-food industry in New Jersey and Pennsylvania. *American Economic Review* 84(4):772–793.

Card, D., and T. Lemieux
1994 *Changing Wage Structure and Black-White Differentials Among Men and Women: A Longitudinal Analysis*. (NBER Working Paper 4755.) Cambridge, MA: National Bureau of Economic Research.
1996 Wage dispersion, returns to skill and black-white wage differentials. *American Economic Review* 74:319–361.

Carr, J.H., and I.F. Megbolugbe
1993 The Federal Reserve Bank of Boston study on mortgage lending revisited. *Journal of Housing Research* 4:277–314.

Case, A., D. Lubotsky, and C. Paxson
2002 Economic status and health in childhood: The origins of the gradient. *American Economic Review* 92(5):1308–1334.

Cavalli-Sforza, L.L.
2000 *Genes, Peoples, and Languages*. New York: North Point Press.

Charles, C.Z.
2001 Processes of residential segregation. In *Urban Inequality: Evidence from Four Cities*, A. O'Connor, C. Tilly, and L. Bobo, eds. New York: Russell Sage Foundation.

Charles, K., and E. Hurst
2002 The transition to home ownership and the black-white wealth gap. *Review of Economics and Statistics* 84(2):281–297.

Chay, K.Y.
1998 The impact of federal civil rights policy on black economic progress: Evidence from the Equal Opportunity Act of 1972. *Industrial and Labor Relations Review* 51(4):608–632.

Chay, K.Y., and M. Greenstone
2000 The convergence in black-white infant mortality rates during the 1960's. *American Economic Review* 90(2):326–332.

Chen, M., and J.A. Bargh
1997 Nonconscious behavioral confirmation processes: The self-fulfilling nature of automatically-activated stereotypes. *Journal of Experimental Social Psychology* 33:541–560.

Chin, M.H., J.X. Zhang, and K. Merrell
1998 Diabetes in the African-American Medicare population: Morbidity, quality of care, and resource utilization. *Diabetes Care* 21(7):1090–1095.

Choy, S.P.
2002 *Access and Persistence: Findings from 10 Years of Longitudinal Research on Students*. Washington, DC: American Council on Education.

Coate, S., and G.C. Loury
1993 Will affirmative-action policies eliminate negative stereotypes? *American Economic Review* 83(5):1220–1240.
Cochran, W.G.
1965 The planning of observational studies of human populations (with discussion). *Journal of the Royal Statistical Society, Series A* 128:234–266.
Coleman, M.G., W.A. Darity, Jr., and R.V. Sharpe
2002 Are Reports of Discrimination Valid? Considering the Moral Hazard Effect. Unpublished paper, Pennsylvania State University, State College.
Collins, J.W., Jr., R.J. David, R. Symons, A. Handler, S.N. Wall, and L. Dwyer
2000 Low-income, African-American mother's perceptions of exposure to racial discrimination and infant birth weight. *Epidemiology* 11:337–339.
Conley, D.
1999 *Being Black, Living in the Red: Race, Wealth and Social Policy in America.* Berkeley: University of California Press.
Cook, S.W., and M. Pelfrey
1985 Reactions to being helped in cooperating interracial groups: A context effect. *Journal of Personality and Social Psychology* 49:1231–1245.
Cook, P.J., and J.H. Laub
1998 The unprecedented epidemic in youth violence. In *Youth Violence, Crime and Justice*, Vol. 24, M. Tonry and M. H. Moore, eds. Chicago: University of Chicago Press.
Cordner, G., B. Williams, and M. Zuniga
2000 *Vehicle Stop Study: Mid-Year Report.* San Diego, CA: San Diego Police Department.
Correll, J., B. Park, C.M. Judd, and B. Wittenbrink
2002 The police officer's dilemma: Using ethnicity to disambiguate potentially threatening individuals. *Journal of Personality and Social Psychology* 83(6):1314–1329.
Council of Economic Advisors
1998 *Changing America: Indicators of Social and Economic Well-Being by Race and Hispanic Origin.* Washington, DC: U.S. Government Printing Office.
Cox, S.M., S.E. Pease, D.S. Miller, and C.B. Tyson
2001 *Interim Report of Traffic Stops Statistics for the State of Connecticut.* Rocky Hill, CT: Division of Criminal Justice.
Crosby, F.
1984 The denial of personal discrimination. *American Behavioral Scientist* 27:371–386.
Crosby, F., S. Bromley, and L. Saxe
1980 Recent unobtrusive studies of black and white discrimination and prejudice: A literature review. *Psychological Bulletin* 87:546–563.
Cross, H., G. Kenney, J. Mell, and W. Zimmerman
1990 *Employer Hiring Practices: Differential Treatment of Hispanic and Anglo Job Seekers.* Washington, DC: Urban Institute Press.
Crow, J.F.
2002 Unequal by nature: A geneticist's perspective on human differences. *Daedalus* (Winter):81–88.
Cunningham, W.A., K.J. Preacher, and M.R. Banaji
2001 Implicit attitude measures: Consistency, stability, and convergent validity. *Psychological Science* 121(2):163–170.
Darity, Jr., W.A.
2003 Employment discrimination, segregation, and health. *American Journal of Public Health* 93(2):226–231.

Darity, W.A., Jr., and P.L. Mason
1998 Evidence on discrimination in employment: Codes of color, codes of gender. *Journal of Economic Perspectives* 12:63–90.
Darity, W.A., Jr., J. Dietrich, and D.K. Guilkey
2001 Persistent advantage or disadvantage? Evidence in support of the intergenerational drag hypothesis. *American Journal of Economics and Sociology* 60(2):435–470.
Darley, J.M., and R.H. Fazio
1980 Expectancy confirmation processes arising in the social interaction sequence. *American Psychologist* 35(10):867–881.
Darley, J.M., and P.H. Gross
1983 A hypothesis-confirming bias in labeling effects. *Journal of Personality and Social Psychology* 44:20–33.
Davis, J.A., T.W. Smith, and P.V. Marsden
2001 *General Social Survey, 1972–2000: Cumulative Codebook*. Chicago: National Opinion Research Center.
Dawid, A.P.
2000 Causal inference without counterfactuals. *Journal of the American Statistical Association* 95:407–448.
Deaton, A.
2003 Health, income, and inequality. *NBER Reporter* (Spring):9–12.
Degler, C.N.
1971 *Neither Black nor White, Slavery and Race Relations in Brazil and the United States*. New York: Macmillan.
de la Garza, R.L., L. DeSipio, F. Garcia, J. Garcia, and A. Falcon
1992 *Latino Voices: Mexican, Puerto Rican and Cuban Perspectives on American Politics*. Boulder, CO: Westview Press.
Del Conte, A., and J. Kling
2001 A synthesis of MTO research on self-sufficiency, safety and health, and behavior and delinquency. *Quarterly Newsletter of the Northwestern University/University of Chicago Joint Center for Poverty Research* 15(1). Available: http://www.jcpr.org/newsletters/vol5_no1/index.html [January 29, 2004].
del Pinal, J.
2003 *Race and Ethnicity in Census 2000*. (Census 2000 Testing, Experimentation, and Evaluation Program, Topic Report 9.) Washington, DC: U.S. Census Bureau.
Denton, N.A., and D.S. Massey
1989 Racial identity among Caribbean Hispanics: The effect of double minority status on residential segregation. *American Sociological Review* 54:790–808.
Dertke, M.C., L.A. Penner, and K. Ulrich
1974 Observer's reporting of shoplifting as a function of thief's race and sex. *Journal of Social Psychology* 94:213–221.
Devine, P.G.
1989 Stereotypes and prejudice: Their automatic and controlled components. *Journal of Personality and Social Psychology* 56:5–18.
2001 Implicit prejudice and stereotyping: How automatic are they? Introduction to a special session. *Journal of Personality and Social Psychology* 81:757–759.
DiNardo, J., N. Fortin, and T. Lemieux
1996 Labor market institutions and the distribution of wages, 1973–1992: A semi-parametric approach. *Econometrica* 64:1001–1044.
Dion, K.L.
2001 Immigrants' perceptions of housing discrimination in Toronto: The housing new Canadians project. *Journal of Social Issues* 57(Fall):523–540.

Donnerstein, E., M. Donnerstein, and C. Koch
1975 Racial discrimination in apartment rentals: A replication. *Journal of Social Psychology* 96:37–38.
Donohue, J.J., J. Heckman, and P. Todd
2002 The schooling of southern blacks: The roles of legal activism and private philanthropy. *Quarterly Journal of Economics* 117(1):225–268.
Dovidio, J.F.
1993 The subtlety of racism. *Training and Development* 47:51–57.
Dovidio, J.F., and S.L. Gaertner, eds.
1986 *Prejudice, Discrimination, and Racism.* San Diego, CA: Academic Press.
Dovidio, J.F., N. Evans, and R.B. Tyler
1986 Racial stereotypes: The contents of their cognitive representations. *Journal of Experimental Social Psychology* 22:22–37.
Dovidio, J.F., J.C. Brigham, B.T. Johnson, and S.L. Gaertner
1996 Stereotyping, prejudice, and discrimination: Another look. In *Stereotypes and Stereotyping*, C.N. Macrae, C. Stangor, and M. Hewstone, eds. New York: Guilford Press.
Dovidio, J.F., K. Kawakami, C. Johnson, B. Johnson, and A. Howard
1997 On the nature of prejudice: Automatic and controlled processes. *Journal of Experimental Social Psychology* 33(5):510–540.
Dovidio, J.F., K. Kawakami, and S.L. Gaertner
2002 Implicit and explicit prejudice and interracial interaction. *Journal of Personality and Social Psychology* 82:62–68.
Duckitt, J.
2001 A dual-process cognitive-motivational theory of ideology and prejudice. In *Advances in Experimental Social Psychology*, Vol. 33, M.P. Zanna, ed. San Diego, CA: Academic Press.
Duncan, O.D.
1968 Inheritance of poverty or inheritance of race? In *On Understanding Poverty*, D. Moynihan, ed. New York: Basic Books.
Duncan, G., and K. Magnuson
2001 Off with Hollingshead: Socioeconomic resources, parenting, and child development. In *Socioeconomic Status, Parenting, and Child Development*, M. Bornstein and R. Bradley, eds. Mahwah, NJ: Lawrence Erlbaum.
2002 Policies to Promote the Health Development of Infants and Preschoolers. (Brookings Policy Paper.) Washington, DC: Brookings Institution Press.
Eagly, A., and S. Chaiken
1993 *Psychology of Attitudes.* Fort Worth, TX: Harcourt Brace Jovanovich.
Elder, G.H.
1974 *Children of the Great Depression: Social Change in Life Experience.* Chicago: University of Chicago Press.
1975 Adolescence in the life cycle. In *Adolescence in the Life Cycle*, S.E. Dragastin and G.H. Elder, Jr., eds. Washington, DC: Hemisphere/Halsted Press.
1985 *Life Course Dynamics: Trajectories and Transitions, 1968–1980.* Ithaca, NY: Cornell University Press.
1991 Lives and social change. In *Status Passages and the Life Course, Volume 1: Theoretical Advances in Life Course Research*, W.R. Heinz, ed. Weinheim, Germany: Deutscher Studien Verlag.
1998 The life course and human development. In *Handbook of Child Psychology*, Vol. 1, R.M. Lerner, ed. New York: John Wiley & Sons.

Emerson, M.O., and C. Smith
2000 *Divided by Faith: Evangelical Religion and the Problem of Race in America*. New York: Oxford University Press.

Emerson, M.O., G. Yancey, and K.J. Chai
2001 Does race matter in residential segregation? Exploring the preferences of white Americans. *American Sociological Review* 66:922–935.

Engel, R.S., and J.M. Calnon
In press Comparing baseline methodologies for police-citizen contacts: Traffic stop data collection for the Pennsylvania state police. *Police Quarterly*.
2001 Examining the Influence of Drivers' Race and Ethnicity During Traffic Stops with Police: Results from a National Survey. (Under review.) Division of Criminal Justice, University of Cincinnati, Cincinnati, OH.

Engel, R.S, J.M Calnon, and T.J. Bernard
2002 Theory and racial profiling: Shortcomings and future directions in research. *Justice Quarterly* 19(2):249–273.

Entwisle, D., and K. Alexander
1992 Summer setback: Race, poverty, school composition, and mathematics achievement in the first two years of school. *American Sociological Review* 57:72–84.
1994 Winter setback: School racial composition and learning to read. *American Sociological Review* 59:446–460.

Entwisle, D., K.L. Alexander, and L.S. Olson
1997 *Children, Schools, and Inequality*. Boulder, CO: Westview Press.

Epstein, A.M., J.Z. Ayanian, J.H. Keogh, S.J. Noonan, N. Armistead, P.D. Cleary, J.S. Weissman, J.A. David-Kasdan, D. Carlson, J. Fuller, D. March, and R. Conti
2000 Racial disparities in access to renal transplantation. *New England Journal of Medicine* 343(21):1537–1544.

Esmail, A., and S. Everington
1993 Racial discrimination against doctors from ethnic minorities. *British Medical Journal* 306(March):691–692.

Essed, P.
1991 *Understanding Everyday Racism*. Thousand Oaks, CA: Sage.
1997 Racial intimidation: Sociopolitical implications of the usage of racist slurs. In *The Language and Politics of Exclusion: Others in Discourse*, S. Riggins and E. Harold, eds. Thousand Oaks, CA: Sage.

Evans, A.S., Jr.
2001 Hate Crime in America: A Sociological Perspective. Paper presented at the Annual Meeting of the Southern Sociological Society, Atlanta, GA, April.

Everett, R.S., and R.A. Wojtkiewicz
2002 Difference, disparity, and race/ethnic bias in federal sentencing. *Journal of Quantitative Criminology* 18(2):189–211.

Fagan, J.
2002 Law, social science, and racial profiling. *Justice Research and Policy* 4(Special Issue).

Farkas, G.
2003 Racial disparities and discrimination in education: What do we know, how do we know it, and what do we need to know? *Teachers College Record* 105(6):1119–1146.

Farley, R.
1990 Blacks, Hispanics, and white ethnic groups: Are blacks uniquely disadvantaged? *American Economic Review* 80(2):237–241.

1996 *The New American Reality: Who We Are, How We Got Here, Where We Are Going.* New York: Russell Sage Foundation.

Farmer, J.J., Jr.
2001 *Monitors' Quarterly and Training Evaluation Report Track State Police Progress: Attorney General Releases Traffic Survey and Semiannual Data on Statewide State Police Traffic Enforcement and Trooper Conduct.* Trenton: New Jersey Department of Law and Public Safety.

Fazio, R.H., and M.A. Olson
2003 Implicit measures in social cognition research: Their meaning and uses. *Annual Review of Psychology* 54:297–327.

Fazio, R.H., J.R. Jackson, B.C. Dunton, and C.J. Williams
1995 Variability in automatic activation as an unobtrusive measure of racial attitudes: A bona fide pipeline? *Journal of Personality and Social Psychology* 69:1013–1027.

Feagin, J.R.
1991 The continuing significance of race: Anti-black discrimination in public places. *American Sociological Review* 56:101–116.

Feagin, J.R., and C.B. Feagin
1996 *Racial and Ethnic Relations*, 5th ed. Upper Saddle River, NJ: Prentice-Hall.

Feagin, J.R., and M.P. Sikes
1994 *Living with Racism: The Black Middle-Class Experience.* Boston, MA: Beacon Press.

Ferguson, R.F.
1998 Can schools narrow the black–white test score gap? In *The Black–White Test Score Gap*, C. Jencks and M. Phillips, eds. Washington, DC: Brookings Institution Press.

Fernandez, R.M.
1997 Spatial mismatch: Housing, transportation, and employment in regional perspective. In *The Urban Crisis: Linking Research to Action*, B. Weisbrod and J. Worthy, eds. Evanston, IL: Northwestern University Press.

Fernandez, R., and L. Sosa
2003 Gendering the Job: Networks and Recruitment at a Call Center. Unpublished manuscript, MIT Sloan School of Management, Cambridge, MA.

Fernandez, R.M., and N. Weinberg
1997 Sifting and sorting: Personal contacts and hiring in a retail bank. *American Sociological Review* 62(December):883–902.

Fernandez, R.M., E. Castilla, and P. Moore
2000 Social capital at work: Networks and employment at a phone center. *American Journal of Sociology* 105(5):1288–1356.

Fischer C.S., M. Hout, M.S. Jankowski, S.R. Lucas, A. Swidler, and K. Voss
1996 *Inequality by Design: Cracking the Myth of the Bell Curve.* Princeton, NJ: Princeton University Press.

Fisher, R.A.
1935 The logic of inductive inference. *Journal of the Royal Statistical Society Series A* 98:39–54.

Fiske, S.T.
1998 Stereotyping, prejudice, and discrimination. In *The Handbook of Social Psychology*, 4th ed., D. Gilbert, S.T. Fiske, and G. Lindzey, eds. New York: McGraw-Hill.
2000 Stereotyping, prejudice, and discrimination at the seam between centuries: Evolution, culture, mind, and brain. *European Journal of Social Psychology* 30:299–322.
2002 What we know now about bias and intergroup conflict: Problem of the century. *Current Directions in Psychological Science* 11:123–128.

Fiske, S.T., and S.E. Taylor
1991 *Social Cognition.* New York: McGraw-Hill.

Fiske, S.T., A.J. Cuddy, P. Glick, and J. Xu
2002 A model of (often mixed) stereotype content: Competence and warmth respectively follow from perceived status and competition. *Journal of Personality and Social Psychology* 82:878–902.

Fix, M., and M.A. Turner
1998 *A National Report Card on Discrimination in America: The Role of Testing.* Washington, DC: Urban Institute Press.

Fix, M., G. Galster, and R. Struyk
1993 An overview of auditing for discrimination. In *Clear and Convincing Evidence: Measurement of Discrimination in America*, M. Fix and R.J. Struyk, eds. Washington, DC: Urban Institute Press.

Fowler, F.J.
1993 *Survey Research Methods*, 2nd edition. Thousand Oaks, CA: Sage.

Fredrickson, G.M.
2002 *Racism: A Short History.* Princeton, NJ: Princeton University Press.

Freedman, D.
2000 From Association to Causation: Some Remarks on the History of Statistics. Technical Report No. 521, Department of Statistics, University of California, Berkeley.
2003 On Specifying Graphical Models for Causation and the Identification Problem. Technical Report No. 601, Department of Statistics, University of California, Berkeley.

Freeman, R.
1973 Changes in the labor market for black Americans, 1948–72. *Brookings Paper on Economic Activity* 1:67–120.
1991 *Crime and the Employment of Disadvantaged Youth.* (NBER Working Paper 3875.) Cambridge, MA: National Bureau of Economic Research.

Gaertner, S.L., and L. Bickman
1971 Effects of race on the elicitation of helping behavior: The wrong number technique. *Journal of Personality and Social Psychology* 20:218–222.

Gaertner, S.L., and J.P. McLaughlin
1983 Racial stereotypes: Associations and ascriptions of positive and negative characteristics. *Social Psychology Quarterly* 46:23–30.

Gail, M.H.
1996 Statistics in action. *Journal of the American Statistical Association* 433:1–13.

Gallup Poll
1999 Racial Profiling Is Seen as Widespread, Particularly Among Young Black Men. Available: http://www.gallup.com/poll/releases/pr991209.asp [January 29, 2004].

Galster, G.C.
1990a Racial discrimination in housing markets during the 1980s: A review of the audit evidence. *Journal of Planning Education and Research* 9(3):165–175.
1990b Racial steering by real estate agents: Mechanisms and motives. *Review of Black Political Economy* 19:39–63.
1998 *An Econometric Model of the Urban Opportunity Structure: Cumulative Causation Among City Markets, Social Problems, and Underserved Areas.* Washington, DC: Fannie Mae Foundation.

Gamoran, A., and R.D. Mare
1989 Secondary school tracking and educational inequality: Compensation, reinforcement, or neutrality? *American Journal of Sociology* 94:1146–1183.

Garrett, B.
2001 Remedying racial profiling. *Columbia Human Rights Law Review* 33:41.

Gary, L.E.
1995 African American men's perceptions of racial discrimination: A sociocultural analysis. *Social Work Research* 19:207–217.

Gilens, M.
1996 "Race coding" and white opposition to welfare. *American Political Science Review* 90:593–604.

Glaser, J.
2003 The Efficacy and Effect of Racial Profiling. Unpublished manuscript. Goldman School of Public Policy, University of California, Berkeley.

Glaser, J., J. Dixit, and D.P. Green
2002 Studying hate crime with the Internet: What makes racists advocate racial violence? *Journal of Social Issues* 58(1):177–194.

Glick, P., and S.T. Fiske
1996 The ambivalent sexism inventory: Differentiating hostile and benevolent sexism. *Journal of Personality and Social Psychology* 70:491–512.

Goldin, C., and C. Rouse
2000 Orchestrating impartiality: The impact of "blind" auditions on female musicians. *American Economic Review* 90(4):715–741.

Goldsmith, A.H., J.R. Veum, and W. Darity, Jr.
2000 Motivation and labor market outcomes. *Worker Well-Being* 19:109–146.

Gomez, J.P., and S.J. Trierweiler
2001 Does discrimination terminology create response bias in questionnaire studies of discrimination? *Personality and Social Psychology Bulletin* 27:630–638.

Goodman, L.A.
1961 Snowball sampling. *Annals of Mathematical Statistics* 20:572–579.

Gordon, M.M.
1964 *Assimilation in American Life: The Role of Race, Religion, and National Origins.* New York: Oxford University Press.

Graham, R.
1990 *The Idea of Race in Latin America 1870–1940.* Austin: University of Texas Press.

Green, D.P., D.Z. Strolovitch, and J.S. Wong
1998 Defended neighborhoods, integration, and racially motivated crime. *American Journal of Sociology* 104(2):372–403.

Green, D.P., R.P. Abelson, and M. Garnett
1999 The distinctive political views of hate-crime perpetrators and white supremacists. In *Cultural Divides: Understanding and Overcoming Group Conflict*, D.A. Prentice and D.T. Miller, eds. New York: Russell Sage Foundation.

Green, D.P., L.H. McFalls, and J.K. Smith
2001 Hate crime: An emergent research agenda. *Annual Review of Sociology* 27:479–504.

Greenwald, A.G., and M.R. Banaji
1995 Implicit social cognition: Attitudes, self-esteem, and stereotypes. *Psychological Review* 102(1):4–27.

Greenwald, A.G., D.E. McGhee, and J.L.K. Schwartz
1998 Measuring individual differences in implicit cognition: The implicit association test. *Journal of Personality and Social Psychology* 74(6):1464–1480.

Griffin, J.H.
1996 *Black Like Me.* New York: Signet.

Guryan, J.
2001 *Desegregation and Black Dropout Rates.* (NBER Working Paper 8345.) Cambridge, MA: National Bureau of Economic Research.

Hagan, J.
1993 The social embeddedness of crime and unemployment. *Criminology* 31:465–491.
Hallinan, M.T.
1998 Diversity effects on student outcomes: Social science evidence. *Ohio State Law Journal* 59:733–754.
Hannaford, I.
1996 *Race: The History of an Idea in the West.* Baltimore, MD: Johns Hopkins University Press.
Harrell, J.P., S. Hall, and J. Taliaferro
2003 Physiological responses to racism and discrimination: An assessment of the evidence. *American Journal of Public Health* 93(2):243–248.
Harrell, S.P.
2000 A multidimensional conceptualization of racism-related stress: Implications for well-being of people of color. *American Journal of Orthopsychiatry* 70:42–57.
Harris, D.A.
1997 "Driving while black" and all other traffic offenses: The Supreme Court and pretextual traffic stops. *Journal of Criminal Law and Criminology* 87(2):544–582.
1999a *Driving While Black: Racial Profiling on Our Nation's Highways.* (An American Civil Liberties Union Special Report.) Washington, DC: American Civil Liberties Union.
1999b The stories, the statistics, and the law: Why "driving while black" matters. *Minnesota Law Review* 84:265–326.
Harris, D.R.
2002 Racial Classification and the 2000 Census. Unpublished paper prepared for the National Research Council Panel to Review the 2000 Census, Washington, DC. Institute for Social Research, University of Michigan.
Harrison, G.W.
1998 Mortgage lending in Boston: A reconsideration of the evidence. *Economic Inquiry* 36(1):29–38.
Harrison, R.J.
2002 Inadequacies of multiple race response data in the federal statistical system. In *Mixed Results*, J. Perlman and M. Waters, eds. New York: Russell Sage Foundation.
Hart, A.J., P.J. Whalen, L.M. Shin, S.C. McInerney, H. Fischer, and S.L. Rauch
2000 Differential response in the human amygdala to racial outgroups vs. ingroup face stimuli. *Neuroreport for Rapid Communication of Neuroscience Research* 11:2351–2355.
Hauser, R.M., S.J. Simmons, and D.I. Pager
2002 High School Dropout, Race-Ethnicity, and Social Background from 1970s to the 1990s. CDE Working Paper No. 2000-12. Center for Demography and Ecology, University of Wisconsin-Madison.
Heckman, J.J.
1979 Sample selection bias as a specification error. *Econometrica* 47:153–161.
1998 Detecting discrimination. *Journal of Economic Perspectives* 12:101–116.
Heckman, J.J., and B.S. Payner
1989 Determining the impact of federal anti-discrimination policy on the economic status of blacks: A study of South Carolina. *American Economic Review* 79(1):138–177.
Heckman, J.J., and P. Siegelman
1993 The Urban Institute audit studies: Their methods and findings. In *Clear and Convincing Evidence: Measurement of Discrimination in America*, M. Fix and R.J. Struyk, eds. Washington, DC: Urban Institute Press.

Hellerstein, J.K., and D. Neumark

1998 Wage discrimination, segregation, and sex differences in wages and productivity within and between plants. *Industrial Relations* 37(2):232–260.

1999 Sex, wages, and productivity: An empirical analysis of Israeli firm-level data. *International Economic Review* 40(1):95–123.

Henderson-King, E.I., and R.E. Nisbett

1996 Anti-black prejudice as a function of exposure to the negative behavior of a single black person. *Journal of Personality and Social Psychology* 1996:654–664.

Henry J. Kaiser Family Foundation

1999 *Race Ethnicity and Medical Care: A Survey of Public Perceptions and Experiences.* Menlo Park, CA: Henry J. Kaiser Family Foundation.

Hewstone, M., M. Rubin, and H. Willis

2002 Intergroup bias. *Annual Review of Psychology* 53:575–604.

Heyns, B.

1978 *Summer Learning and the Effects of Schooling.* New York: Academic Press.

Higginbotham, A.

1996 *Shades of Freedom: Racial Politics and Presumptions of the American Legal Process.* New York: Oxford University Press.

Hill, A.B.

1987 The environment and disease: Association or causation? In *Evolution of Epidemiologic Ideas: Annotated Readings on Concepts in Methods*, S. Greenland, ed. Newton Lower Falls, MA: Epidemiology Resources Inc.

Hirschman, C., R. Alba, and R. Farley

2000 The meaning and measurement of race in the U.S. census: Glimpses into the future. *Demography* 37(2):381–393.

Hobsbawm, E.

1992 Introduction: Inventing traditions. In *The Invention of Tradition*, E. Hobsbawm and T. Ranger, eds. Cambridge, England: Cambridge University Press.

Holland, P.W.

1986 Statistics and causal inference (with comments). *Journal of the American Statistical Association* 396:945–970.

2003 *Causation and Race.* Research Report 03-03. Princeton, NJ: Educational Testing Service.

Hollinger, D.

2000 *Postethnic America: Beyond Multiculturalism*, 2nd ed. New York: Basic Books.

Holzer, H.J.

1996 *What Employers Want.* New York: Russell Sage Foundation.

Holzer, H.J., and J. Ludwig

2003 Measuring discrimination in education: Are methodologies from labor and housing markets useful? *Teachers College Record* 105(6):1147–1178.

Hsiao, C.

1986 *Analysis of Panel Data.* Cambridge, England: Cambridge University Press.

Huddy, L., and S. Virtanen

1995 Subgroup differentiation and subgroup bias among Latinos as a function of familiarity and positive distinctiveness. *Journal of Personality and Social Psychology* 68:97–108.

Hughes, D., and L. Chen

1999 Parents' race-related messages to children: A developmental perspective. In *Child Psychology: A Handbook of Contemporary Issues*, C. Tamis-Lemonda and L. Balter, eds. New York: New York University Press.

Hurh, W.M., and K.C. Kim

1989 The "success" image of Asian Americans: Its validity and its practical and theoretical implications. *Ethnic and Racial Studies* 12(4):512–537.

Ichioka, Y.

1977 The early Japanese quest for citizenship: The background of the 1922 Ozawa case. *Amerasia Journal* 4:1–22.

Ignatiev, N.

1995 *How the Irish Became White.* New York: Routledge.

Ihlanfeldt, K.R., and D.L. Sjoquist

1998 The spatial mismatch hypothesis: A review of recent studies and their implication for welfare reform. *Housing Policy Debate* 9(4):849–892.

Imperato, P.J., R.P. Nenner, and T.O. Will

1996 Radical prostatectomy: Lower rates among African American men. *Journal of the National Medical Association* 88(9):589–594.

Institute of Medicine

2003 *Unequal Treatment: Confronting Racial and Ethnic Disparities in Health Care,* B.D. Smedley, A.Y. Stith, and A.R. Nelson, eds. Committee on Understanding and Eliminating Racial and Ethnic Disparities in Health Care, Board on Health Sciences Policy. Washington, DC: The National Academies Press.

Institute on Race and Poverty

2001 *Components of Racial Profiling Legislation.* Minneapolis: University of Minnesota Law School.

Jackman, M.R.

1994 *The Velvet Glove: Paternalism and Conflict in Gender, Class, and Race Relations.* Berkeley: University of California Press.

Jacobson, J., C. Olsen, J.K. Rice, S. Sweetland, and J. Ralph

2001 *Educational Achievement and Black-White Inequality.* (NCES 2001-061.) Washington, DC: National Center for Education Statistics.

Jencks, C., and S. Mayer

1990 The social consequences of growing up in a poor neighborhood. In *Inner-City Poverty in the United States,* L. Lynn and M. McGeary, eds. Committee on National Urban Policy, National Research Council. Washington, DC: National Academy Press.

Johnson, G.E., and F.P. Stafford

1998 Alternative approaches to occupational exclusion. In *Women's Work and Wages,* I. Persson and C. Jonung, eds. London: Routledge.

Johnson, W.R., and D.A. Neal

1998 Black skills and the black-white earnings gap. In *The Black-White Test Score Gap,* C. Jencks and M. Phillips, eds. Washington, DC: Brookings Institution Press.

Jones, F.L.

1983 On decomposing the wage gap: A critical comment on Blinder's method. *Journal of Human Resources* 18:126–130.

Jones, J.M.

1997 *Prejudice and Racism,* 2nd ed. New York: McGraw-Hill.

Judd, C.M., B. Park, C.S. Ryan, M. Brauer, and S. Kraus

1995 Stereotypes and ethnocentrism: Diverging interethnic perceptions of African American and white American youth. *Journal of Personality and Social Psychology* 69(3):460–481.

Juhn, C., K.M. Murphy, and B. Pierce

1991 Accounting for the slowdown in black-white wage convergence. In *Workers and*

Their Wages: Changing Patterns in the United States, M.H. Kosters, ed. Washington, DC: AEI Press.

Jussim, L.

1989 Teacher expectations: Self-fulfilling prophecies, perceptual biases, and accuracy. *Journal of Personality and Social Psychology* 57(3):469–480.

1991 Social perception and social reality: A reflection-construction model. *Psychological Review* 98:54–73.

Jussim, L., and J.S. Eccles

1992 Teacher expectations II: Construction and reflection of student achievement. *Journal of Personality and Social Psychology* 63(6):947–961.

Jussim, L., J. Eccles, and S.J. Madon

1996 Social perception, social stereotypes, and teacher expectations: Accuracy and the quest for the powerful self-fulfilling prophecy. *Advances in Experimental Social Psychology* 29:281–388.

Kahlenberg, R.

2001 An unambitious legacy. *Education Week*. February 21, p. 48.

Kahn, L.M.

1991 Discrimination in professional sports: A survey of the literature. *Industrial and Labor Relations Review* 44:395–418.

Kahn, L.M., and P.D. Sherer

1988 Racial differences in professional basketball players' compensation. *Journal of Labor Economics* 6(1):40–61.

Kain, J.F.

1968 Housing segregation, Negro employment, and metropolitan decentralization. *Quarterly Journal of Economics* 82(2):175–197.

Kasarda, J., and K. Ting

1996 Joblessness and poverty in America's central cities: Causes and policy prescriptions. *Housing Policy Debate* 7(2):387–419.

Katz, I., and R.G. Hass

1988 Racial ambivalence and American value conflict: Correlational and priming studies of dual cognitive structures. *Journal of Personality and Social Psychology* 55:893–905.

Katz, I., J. Wackenhut, and R.G. Hass

1986 Racial ambivalence, value duality, and behavior. In *Prejudice, Discrimination, and Racism*, J.F. Dovidio and S.L. Gaertner, eds. San Diego, CA: Academic Press.

Kawakami, K., and J.F. Dovidio

2001 The reliability of implicit stereotyping. *Personality and Social Psychology Bulletin* 27(2):212–225.

Kawakami, K., K.L. Dion, and J.F. Dovidio

1998 Racial prejudice and stereotype activation. *Personality and Social Psychology Bulletin* 24(4):407–416.

Keith, V.M., and C. Herring

1991 Skin tone and stratification in the black community. *American Journal of Sociology* 97(3):760–778.

Keppel, K.G., J.N. Pearcy, and D.K. Wagener

2002 Trends in racial and ethnic-specific rates for the health status indicators: United States, 1990–1998. *Healthy People 2000 Statistical Notes*. 23(January):1–16.

Kerckhoff, A.C.

1986 Effects of ability grouping in British secondary schools. *American Sociological Review* 51(6):842–858.

Kinder, D.R.
1998 Opinion and action in the realm of politics. In *Handbook of Social Psychology*, 4th ed., D.T. Gilbert, S.T. Fiske, and G. Lindzey, eds. New York: McGraw-Hill.

Kinder, D., and D. Kiewiet
1981 Sociotropic politics: The American case. *British Journal of Politics* 11:129–161.

Kinder, D.R., and L.M. Sanders
1996 *Divided by Color: Racial Politics and Democratic Ideals.* Chicago: University of Chicago Press.

Kinder, D.R., and D.O. Sears
1981 Prejudice and politics: Symbolic racism versus racial threats to the good life. *Journal of Personality and Social Psychology* 40:414–431.

Kirschenman, J., and K.M. Neckerman
1991 "We'd love to hire them, but . . ." The meaning of race for employers. In *The Urban Underclass*, C. Jencks and P.E. Peterson, eds. Washington, DC: Brookings Institution Press.

Kitano, H.H., and S. Sue
1973 The model minorities. *Journal of Social Issues* 29(2):1–10.

Klein, H.S.
1999 *The Atlantic Slave Trade.* New York: Cambridge University Press.

Kluegel, J.R., and L.D. Bobo
2001 Perceived group discrimination and policy attitudes: The sources and consequences of the race and gender gaps. In *Urban Inequality: Evidence from Four Cities*, A. O'Conner, C. Tilly, and L.D. Bobo, eds. New York: Russell Sage Foundation.

Kocieniewski, D.
2002 New Jersey troopers avoid jail in case that highlighted profiling. *The New York Times*, January 15.

Kornhaber, M.
1997 Seeking Strengths: Equitable Identification for Gifted Education and the Theory of Multiple Intelligences. Doctoral dissertation, Harvard Graduate School of Education, Cambridge, MA.

Krieger, N.
1990 Racial and gender discrimination: Risk factors for high blood pressure? *Social Science and Medicine* 39:1273–1281.
1994 Epidemiology and the web of causation: Has anyone seen the spider? *Social Science and Medicine* 39:887–903.
1999 Embodying inequality: A review of concepts, measures, and methods for studying health consequences of discrimination. *International Journal of Health Services* 29(2):295–352.
2000 Discrimination and health. In *Social Epidemiology*, L.F. Berkman and I. Kawachi, eds. New York: Oxford University Press.
2001 A glossary for social epidemiology. *Journal of Epidemiology and Community Health* 55:693–700.
2003 Does racism harm health? Did child abuse exist before 1962? On explicit questions, critical science, and current controversies: An ecosocial perspective. *American Journal of Public Health* 93(2):194–199.

Krieger, N., and S. Sidney
1996 Racial discrimination and blood pressure: The CARDIA study of young black and white adults. *American Journal of Public Health* 86:1370–1378.

Krueger, A.B.
2002 What's in a name? Perhaps plenty if you're a job seeker. *The New York Times*, December 12.

Kuh, D., and Y. Ben-Shlomo
1997 *A Lifecourse Approach to Chronic Disease Epidemiology: Tracing the Origins of Ill Health from Early to Adult Life.* Oxford: Oxford University Press.
Ladd, H.F.
1998 Evidence on discrimination in mortgage lending. *Journal of Economic Perspectives* 12:41–62.
Lamberth, J.
1994 Revised Statistical Analysis of the Incidence of Police Stops and Arrests of Black Drivers/Travelers on the New Jersey Turnpike Between Exits or Interchanges 1 and 3 from the Years 1988 Through 1991. Unpublished document, Department of Psychology, Temple University, Philadelphia, PA.
1996 Report of John Lamberth, Ph.D., to the American Civil Liberties Union. Available: http://archive.aclu.org/court/lamberth.html [January 29, 2004].
Landrine, H., and E.A. Klonoff
1996 The schedule of racist events: A measure of racial discrimination and a study of its negative physical and mental health consequences. *Journal of Black Psychology* 22:144–168.
Lange, J.E., K.O. Blackman, and M.B. Johnson
2001 *Speed Violation Survey of the New Jersey Turnpike: Final Report.* Trenton, NJ: Office of the Attorney General.
Lansdowne, W.M.
2000 *Vehicle Stop Demographic Study.* San Jose, CA: San Jose Police Department.
Lee, V.E., D.T. Burkam, and L.F. LoGerfo
2001 Who Goes Where? How Kindergartners' Social Background Maps Onto the Quality of the Elementary Schools They Attend. Paper presented at the annual meeting of the American Educational Research Association, Seattle, WA, April.
1990 The impact of affirmative action regulation and equal employment law on black employment. *Journal of Economic Perspectives* 4(4):47–64.
Leventhal, T., and J. Brooks-Gunn
2000 The neighborhoods they live in: The effects of neighborhood residence on child and adolescent outcomes. *Psychological Bulletin* 126(2):309–337.
Levine, J.M., and R.L. Moreland
1998 Small groups. In *The Handbook of Social Psychology*, 4th ed., D. Gilbert, S. Fiske, and G. Lindzey, eds. Boston, MA: McGraw-Hill.
Lieberman, L.
1993 Race and ethnicity: Overlapping meanings. *ANN World Focus* 1(1):3–4.
Lieberman, R.C.
1998 *Shifting the Color Line: Race and the American Welfare State.* Cambridge, MA: Harvard University Press.
Lillie-Blanton, M., P.E. Parson, H. Gayle, and A. Dievler
1996 Racial differences in health: Not just black and white, but shades of gray. *Annual Review of Public Health* 17:411–448.
Linville, P.W., and E.E. Jones
1980 Polarized appraisals of out-group members. *Journal of Personality and Social Psychology* 38:689–703.
Lloyd, K.M, M. Tienda, and A. Zajacova
2002 Trends in the educational achievement of minority students since Brown v. Board of Education. In *Achieving High Educational Standards for All*, T. Ready, C. Edley, Jr., and C.E. Snow, eds. Division of Behavioral and Social Sciences and Education, National Research Council. Washington, DC: National Academy Press.

Lochner, L.J.
1999 *Education, Work, and Crime: Theory and Evidence.* (RCER Working Paper No. 465.) Rochester, NY: Rochester Center for Economic Research.

Loury, G.C.
1977 A dynamic theory of racial income differences. In *Women, Minorities and Employment Discrimination*, P.A. Wallace and A.M. LaMond, eds. Lexington, MA: D.C. Heath and Co.
2002 *The Anatomy of Racial Inequality.* Cambridge, MA: Harvard University Press.

Lucas, S.R.
1994 Effects of Race and Gender Discrimination in the United States: 1940–1980. Unpublished Ph.D. dissertation, University of Wisconsin, Madison.
1999 *Tracking Inequality.* New York: Teachers College Press.

Lucas, S.R., and M. Berends
2002 Sociodemographic diversity, correlated achievement, and de facto tracking. *Sociology of Education* 75(4):328–348.

Lucas, S.R., and A. Gamoran
1991 Race and Track Assignment: A Reconsideration with Course-Based Indicators of Track Locations. Presentation at the American Sociological Association annual meeting, Cincinnati, OH, August.
2002 Tracking and the achievement gap. In *Bridging the Gap*, J.E. Chubb and T. Loveless, eds. Washington, DC: Brookings Institution Press.

Ludwig, J., G.J. Duncan, and P. Hirschfield
2001 Urban poverty and juvenile crime: Evidence from a randomized housing-mobility experiment. *Quarterly Journal of Economics* 116(2):655–680.

Lundberg, S.J.
1991 The enforcement of equal opportunity laws under imperfect information: Affirmative action and alternatives. *Quarterly Journal of Economics* 106(1):309–326.

Lundberg, S.J., and R. Startz
1983 Private discrimination and social intervention in competitive labor markets. *American Economic Review* 73(3):340–347.
1998 On the persistence of racial inequality. *Journal of Labor Economics* 16(2):292–324.
2000 Inequality and race: Models and policy. In *Meritocracy and Economic Inequality*, K. Arrow, S. Bowles, and S. Durlauf, eds. Princeton, NJ: Princeton University Press.

Luttmer, E.F.P.
2001 Group loyalty and the taste for redistribution. *Journal of Political Economy* 109(3):500–528.

Maddox, K.B., and S.A. Gray
2002 Cognitive representations of black Americans: Reexploring the role of skin tone. *Personality and Social Psychology Bulletin* 28(2):250–259.

Manski, C.F.
1995 *Identification Problems in Social Sciences.* Cambridge, MA: Harvard University Press.
2003 *Partial Identification of Probability Distributions.* New York: Springer-Verlag.

Mare, R.D.
1995 Changes in educational attainment and social enrollment. In *State of the Union: America in the 1990s*, Vol. 1, R. Farley, ed. New York: Russell Sage Foundation.

Margo, R.
1990 *Race and Schooling in the South, 1880–1950: An Economic History.* Chicago: University of Chicago Press.

Marini, M.M., and B. Singer
1988 Causality in the social sciences. In *Sociological Methodology*, C.C. Clogg, ed. Washington, DC: American Sociological Association.
Marlow, D., and D.P. Crowne
1961 Social desirability and response to perceived situational demands. *Journal of Consulting Psychology* 25:109–115.
Martell, R.F., D.M. Lane, and C. Emrich
1996 Male-female differences: A computer simulation. *American Psychologist* 51:157–158.
Martin, E., T.J. DeMaio, and P.C. Campanelli
1990 Contest effects for census measure of race and Hispanic origin. *Public Opinion Quarterly* 54:551–566.
Massey, D.S.
2001 Residential segregation and neighborhood conditions in U.S. metropolitan areas. In *America Becoming: Racial Trends and Their Consequences*, Vol. 1, N.J. Smelser, W.J. Wilson, and F. Mitchell, eds. Commission on Behavioral and Social Sciences and Education. Washington, DC: National Academy Press.
Massey, D.S., and N.A. Denton
1992 Racial identity and the segregation of Mexicans in the United States. *Social Science Research* 21:235–260.
1993 *American Apartheid: Segregation and the Making of the Underclass.* Cambridge, MA: Harvard University Press.
Massey, D.S., and G. Lundy
2001 Use of black English and racial discrimination in urban housing markets: New methods and findings. *Urban Affairs Review* 36(4):452–469.
Massey, D.S., C.Z. Charles, and G. Dinwiddie
2003 The Continuing Consequences of Segregation: Family Stress and College Academic Performance. Unpublished paper, Department of Sociology, University of Pennsylvania.
Mayberry, R.M., F. Mili, and E. Ofili
2000 Racial and ethnic differences in access to medical care. *Medical Care Research and Review, 2000 Supplement 1* 57:108–146.
Mayr, E.
2002 The biology of race and the concept of equality. *Daedalus* (Winter):89–94.
Mays, V.M., L.M. Coleman, and J.S. Jackson
1996 Perceived race-based discrimination, employment status, and job stress in a national sample of black women: Implications for health outcomes. *Journal of Occupational Health Psychology* 1(3):319–329.
McConahay, J.B.
1986 Modern racism, ambivalence, and the modern racism scale. In *Prejudice, Discrimination, and Racism*, J.F. Dovidio and S.L. Gaertner, eds. San Diego, CA: Academic Press.
McMahon, L.F., R.A. Wolfe, S. Huang, P. Tedeschi, W. Manning, and M.J. Edlund
1999 Racial and gender variation in use of diagnostic colonic procedures in the Michigan Medicare population. *Medical Care* 37(7):712–717.
Mead, M., T. Dobzhansky, E. Tobach, and R. Light, eds.
1968 *Science and the Concept of Race.* New York: Columbia University Press.
Meyer, B.D.
1995 Natural and quasi-experiments in economics. *Journal of Business and Economic Statistics* 13:151–162.

Mickelson, R.A.
2001 Subverting Swann: First- and second-generation segregation in the Charlotte–Mecklenburg schools. *American Educational Research Journal* 38:215–252.
2003 When are racial disparities in education the result of racial discrimination? A social science perspective. *Teachers College Record* 105(6):1052–1086.
Montagu, A.
1972 *Statement on Race.* New York: Oxford University Press.
Morgan, S.L.
2001 Counterfactuals, causal effect heterogeneity, and the Catholic school effect on learning. *Sociology of Education* 74:341–374.
Morin, A.
2001 Misperceptions cloud whites' view of blacks. *The Washington Post*, July 11.
Mouw, T.
2000 Job relocation and the racial gap in unemployment in Detroit and Chicago, 1980–1990. *American Sociological Review* 65:730–753.
Munnell, A.H., L.E. Browne, J. McEneaney, and G.M.B. Tootell
1992 *Mortgage Lending in Boston: Interpreting HMDA Data.* (Working Paper No. 92-7.) Boston, MA: Federal Reserve Bank of Boston.
Munnell, A.H., G.M.B. Tootell, L.E. Browne, and J. McEneaney
1996 Mortgage lending in Boston: Interpreting HMDA data. *American Economic Review* 86(1):25–53.
Murnane, R.J., J.B. Willett, and F. Levy
1995 *The Growing Importance of Cognitive Skills in Wage Determination.* (NBER Working Paper 5076.) Cambridge, MA: National Bureau of Economic Research.
Murphy, S.
2002 Audit studies and the assessment of discrimination. In *Measuring Housing Discrimination in a National Study: Report of a Workshop*, A.W. Foster, F. Mitchell, and S.E. Fienberg, eds. Division of Behavioral and Social Sciences and Education, National Research Council. Washington, DC: National Academy Press.
Murray, C.B., and J.S. Jackson
1982–1983 The conditioned failure model of black educational underachievement. *Humboldt Journal of Social Relations* 10(1):276–300.
Nagel, J.
1994 Constructing ethnicity: Creating and recreating ethnic identity and culture. *Social Problems* 41(February):153.
National Center for Education Statistics
1996 Racial and Ethnic Classifications Used by Public Schools. NCES 96-002. Available: http://nces.ed.gov/pubs98/98035.pdf [January 29, 2004].
National Center for Health Statistics
2001 *Health, United States, 2001 with Urban and Rural Health Chartbook.* Hyattsville, MD: National Center for Health Statistics.
National Conference of State Legislatures
2002 State Crime Legislation in 2001. Available: http://www.ncsl.org/programs/cj/01crime.htm [January 29, 2004].
National Fair Housing Alliance
2002 *2002 Housing Trends Report.* Washington, DC: National Fair Housing Alliance.
National Research Council
1989 *A Common Destiny: Blacks and American Society*, G.D. Jaynes and R.M. Williams, Jr., eds. Committee on the Status of Black Americans, Commission on Behavioral and Social Sciences and Education. Washington, DC: National Academy Press.
1996 *Spotlight on Heterogeneity: The Federal Standards for Racial and Ethnic Classifica-*

tion. Summary of a Workshop, B. Edmonston, J. Goldstein, and J.T. Lott, eds. Committee on National Statistics, Commission on Behavioral and Social Sciences and Education. Washington, DC: National Academy Press.

1999 *High Stakes: Testing for Tracking, Promotion, and Graduation,* J.P. Heubert and R. Hauser, eds. Committee on Appropriate Test Use, Commission on Behavioral and Social Sciences and Education. Washington, DC: National Academy Press.

2001a *America Becoming: Racial Trends and Their Consequences,* Vol. 1, N.J. Smelser, W.J. Wilson, and F. Mitchell, eds. Commission on Behavioral and Social Sciences and Education. Washington, DC: National Academy Press.

2001b *Evaluating Welfare Reform in an Age of Transition,* R.A. Moffitt and M. Ver Ploeg, eds. Committee on National Statistics. Division on Behavioral and Social Sciences and Education. Washington, DC: National Academy Press.

2002a *Achieving High Educational Standards for All: A Conference Summary,* T. Ready, C. Edley, Jr., and C.E. Snow, eds. Division of Behavioral and Social Sciences and Education. Washington, DC: National Academy Press.

2002b *Measuring Housing Discrimination in a National Study: Report of a Workshop,* A.W. Foster, F. Mitchell, and S.E. Fienberg, eds. Committee on National Statistics, Division of Behavioral and Social Sciences and Education. Washington, DC: National Academy Press.

2004 *The 2000 Census: Counting Under Adversity,* C.F. Citro, D.L. Cork, and J.L. Norwood, eds. Panel to Review the 2000 Census, Committee on National Statistics, Division on Behavioral and Social Sciences and Education. Washington, DC: The National Academies Press.

National Research Council and Institute of Medicine

2001 *Juvenile Crime, Juvenile Justice,* J. McCord, C.S. Widom, and N.A. Crowell, eds. Panel on Juvenile Crime: Prevention, Treatment, and Control, Committee on Law and Justice and Board on Children, Youth, and Families. Washington, DC: National Academy Press.

Natriello, G., E.L. McDill, and A.M. Pallas

1990 *Schooling Disadvantaged Children: Racing Against Catastrophe.* New York: Teachers College Press.

Neal, D.A., and W.R. Johnson

1996 The role of pre-market factors in black-white wage differences. *Journal of Political Economy* 104(5):869–895.

Neckerman, K.M., and J. Kirschenman

1991 Hiring strategies, racial bias, and inner-city workers. *Social Problems* 38:433–437.

Nelson, R.L., and E. Bennett

2003 Judicial Treatment of Statistical Evidence in Cases Alleging Racial Discrimination in Employment: Now (2000–2002) and Then (1980–82). Unpublished paper, Department of Sociology, Northwestern University, Evanston, IL.

Neumark, D.

1996 Sex discrimination in restaurant hiring: An audit study. *Quarterly Journal of Economics* 3(3):915–942.

Neumark, D., and M. McLennan

1995 Sex discrimination and women's labor market outcomes. *Journal of Human Resources* 30(4):713–740.

Neumark, D., and W.A. Stock

2001 *The Effects of Race and Sex Discrimination Laws.* (NBER Working Paper 8111.) Cambridge, MA: National Bureau of Economic Research.

Newman, L.S., and R. Erber, eds.
2002 *Understanding Genocide: The Social Psychology of the Holocaust.* New York: Oxford University Press.

Newport, F.
1999 Racial Profiling Is Seen as Widespread, Particularly Among Young Black Men. Gallup Poll Analyses. Available: www.gallup.com [February 2, 2004].

Newport, F., J. Ludwig, and S. Kearney
2001 Black-White Relations in the United States, 2001 Update. Gallup Poll Social Audit. Available: www.gallup.com [February 2, 2004].

Neyman, J.
1923 Sur les applications de la theorie des probabilities aux experiences agricoles: Essai des principes. *Roczniki Nauk Rolniczki* 10:1–51, in Polish. English translation by D. Dabrowska and T. Speed, 1990. *Statistical Science* 5:463–480.

Nobles, M.
2000 *Shades of Citizenship: Race and the Census in Modern Politics.* Palo Alto, CA: Stanford University Press.

Nopo, H.
2002 Matching as a Tool to Decompose Wage Gaps. Unpublished paper, Department of Economics, Northwestern University, Evanston, IL.

Northrup, D.
1994 *The Atlantic Slave Trade.* Lexington, MA: D.C. Heath and Co.

Oakes, J.
1985 *Keeping Track: How Schools Structure Inequality.* New Haven, CT: Yale University Press.
1994 More than misapplied technology: A normative and political response to Hallinan on tracking. *Sociology of Education* 67:84–88.

Oakes, J., K. Muir, and R. Joseph
2000 Course Taking and Achievement in Math and Science: Inequalities that Endure. Paper presented at the National Institute for Science Education Conference, May, Detroit, MI.

Oaxaca, R.L.
1973 Sex discrimination in wages. In *Discrimination in Labor Markets*, O. Ashenfelter and A. Rees, eds. Princeton, NJ: Princeton University Press.

Oaxaca, R.L., and M.R. Ransom
1999 Identification in detailed wage decompositions. *Review of Economics and Statistics* 81(1):154–157.

O'Callaghan, M.
1980 *Sociological Theories: Race and Colonialism.* Paris: United Nations Educational, Social, and Cultural Organization.

O'Conner, A., C. Tilly, and L.D. Bobo, eds.
2001 *Urban Inequality: Evidence from Four Cities.* New York: Russell Sage Foundation.

Office of Information and Regulatory Affairs
1997 Revisions to the standards for the classification of federal data on race and ethnicity: Federal Register Announcement, Executive Office of the President, Office of Management and Budget. *Federal Register* 62(21):58,782–58,790.

Oliver, M.T., and T.M. Shapiro
1995 *Black Wealth, White Wealth: A New Perspective on Racial Inequality.* New York: Routledge.

Omi, M.A.
2001 The changing meaning of race. In *America Becoming: Racial Trends and Their Consequences*, Vol. 1, N.J. Smelser, W.J. Wilson, and F. Mitchell, eds. Commission on

Behavioral and Social Sciences and Education. Washington, DC: National Academy Press.

Omi, M.A., and H. Winant
1986 *Racial Formation in the United States: From the 1960s to the 1990s*, 2nd ed. New York: Routledge.

Ondrich, J., S.L. Ross, and J. Yinger
2000 How common is housing discrimination? Improving on traditional measures. *Journal of Urban Economics* 47:470–500.

Oppenheimer, D.B.
1993 Negligent discrimination. *University of Pennsylvania Law Review* 141(January):899.

Orfield, G., and S.E. Eaton
1996 *Dismantling Desegregation: The Quiet Reversal of Brown v. Board of Education.* Cambridge, MA: Harvard University Press.

Orne, M.T.
1962 On the social psychology of the psychological experiment: With particular reference to demand characteristics and their implications. *American Psychologist* 17:776–783.

Pablos-Mendez, A.
2001 Brief 5: Tuberculosis. Health and Nutrition Emerging and Reemerging Issues in Developing Countries. Available: http://www.ifpri.org/2020/focus/focus05/focus05_05.htm [January 29, 2004].

Pager, D.
2002 The Mark of a Criminal Record. Available: http://www.northwestern.edu/ipr/publications/papers/2002/WP-02-37.pdf [January 29, 2004].

Park, R.
1950 *Race and Culture.* Glencoe, IL: The Free Press.

Pearl, J.
2000 *Causality: Models, Reasoning, and Inference.* Cambridge, England: Cambridge University Press.

Petersen, T., and L.A. Morgan
1995 Separate and unequal: Occupation-establishment sex segregation and the gender wage gap. *American Journal of Sociology* 101:329–365.

Petersen, T., I. Saporta, and M. Seidel
2000 Offering a job: Meritocracy and social networks. *American Journal of Sociology* 106(3):763–816.

Pettigrew, T.F.
1998a Intergroup contact theory. *Annual Review of Psychology* 49:65–85.
1998b Reactions toward the new minorities of Western Europe. *Annual Review of Sociology* 24:77–103.

Pettigrew, T.F., and R.W. Meertens
1995 Subtle and blatant prejudice in Western Europe. *European Journal of Social Psychology* 25:57–75.

Pettigrew, T.F., and L.R. Tropp
2000 Does intergroup contact reduce prejudice: Recent meta-analytic findings. In *Reducing Prejudice and Discrimination*, S. Oskamp, ed. Mahwah, NJ: Lawrence Erlbaum.

Petty, R.E., and J.A. Krosnick
1995 *Attitude Strength: Antecedents and Consequences.* Mahwah, NJ: Lawrence Erlbaum.

Pfeffer, J.
1998 Understanding organizations: Concepts and controversies. In *Handbook of Social*

Psychology, Vol. 2, 4th ed., D. Gilbert, S. Fiske, and G. Lindzey, eds. New York: McGraw-Hill.

Phelps, E.S.
1972 The statistical theory of racism and sexism. *American Economic Review* 62:659–661.

Phelps, E.A., K.J. O'Connor, W.A. Cunningham, E.S. Funayama, J.C. Gatenby, J.C. Gore, and M.R. Banaji
2000 Performance on indirect measures of race evaluation predicts amygdala activity. *Journal of Cognitive Neuroscience* 12:1–10.

Phillips, M., J. Crouse, and J. Ralph
1998 Does the black-white test score gap widen after children enter school? In *The Black-White Test Score Gap*, C. Jenks and M. Phillips, eds. Washington, DC: Brookings Institution Press.

Plous, S., and T. Williams
1995 Racial stereotypes from the days of American slavery: A continuing legacy. *Journal of Applied Social Psychology* 25:795–817.

Police Foundation
2001 *Racial Profiling: The State of the Law.* Washington, DC: Police Foundation.

Pratt, J.W., and R. Schlaifer
1984 On the nature and discovery of structure (with discussion). *Journal of the American Statistical Association* 79:9–33.
1988 On the interpretation and observation of laws. *Journal of Econometrics* 39:23–52.

Preston, C.
1998 Perceptions of discriminatory practices and attitudes: A survey of African American librarians. *College and Research Libraries* (Sept.):434–445.

Quindlen, A.
2002 Armed with only a neutral lipstick. *Newsweek*, March 18.

Ramirez, D., J. McDevitt, and A. Farrell
2000 *A Resource Guide on Racial Profiling Data Collection Systems: Promising Practices and Lessons Learned.* (NCJ 184768.) Washington, DC: U.S. Department of Justice.

Rees, A., and G.P. Shultz
1970 *Workers and Wages in an Urban Labor Market.* Chicago: University of Chicago Press.

Riach, P.A., and J. Rich
2002 Field experiments of discrimination in the market place. *The Economic Journal* 112(Nov.):F480–F518.

Ridley, S.E., J.A. Bayton, and J.H. Outtz
1989 Taxi Service in the District of Columbia: Is it Influenced by Patron's Race and Destination? Paper prepared for the Washington, D.C., Lawyer's Committee for Civil Rights Under the Law.

Roediger, D.R.
1991 *The Wages of Whiteness: Race and the Making of the American Working Class.* London, England: Verso.

Rojek, J., R. Rosenfeld, and S. Decker
In press The influence of driver's race on traffic stops in Missouri. *Police Quarterly.*

Romero, F.S.
2000 The Supreme Court and the protection of minority rights: An empirical examination of racial discrimination cases. *Law and Society* 24:291–313.

Roper Center for Public Opinion Research
1982 *A Guide to the Roper Center Resources for the Study of American Race Relations.* Storrs, CT: Roper Center.

Rosenbaum, P.

2001 Replicating effects and biases. *American Statistician* 55:223–227.

2002 *Observational Studies*, 2nd edition. New York: Springer-Verlag.

Rosenberg, N., J.K. Pritchard, J.L. Weber, H.M. Cann, K.K. Kidd, L.A. Zhivotovsky, and M. Feldman

2002 Genetic structure of human populations. *Science* 298:2381–2385.

Rosenthal, R.

1976 *Experimenter Effects in Behavioral Research*. New York: John Wiley & Sons.

2002 The Pygmalion effect and its mediating mechanisms. In *Improving Academic Achievement: Impact of Psychological Factors on Education*, J. Aronson, ed. San Diego, CA: Academic Press.

Ross, S.L.

2002 Paired testing and the 2000 Housing Discrimination Survey. In National Research Council, *Measuring Housing Discrimination in a National Study: Report of a Workshop*, A.W. Foster, F. Mitchell, and S.E. Fienberg, eds., Committee on National Statistics, Division of Behavioral and Social Sciences and Education. Washington, DC: National Academy Press.

Ross, S.L., and J. Yinger

2002 Detecting Discrimination: A Comparison of the Methods Used by Scholars and Civil Rights Enforcement Officials. Unpublished paper, Department of Economics, University of Connecticut, Storrs.

Rubin, D.

1974 Estimating causal effects of treatments in randomized and nonrandomized studies. *Journal of Educational Psychology* 66:688–701.

1976 Inference and missing data. *Biometrica* 63(3):581–592.

1977 Assignment of treatment group on basis of a covariate. *Journal of Educational Statistics* 2:1–26.

1978 Bayesian inference for causal effects: The role of randomization. *Annals of Statistics* 6:34–58.

Rudman, L.A., A.G. Greenwald, D.S. Mellott, and J.L.K. Schwartz

1999 Measuring the automatic components of prejudice: Flexibility and generality of the Implicit Association Test. *Social Cognition* 17(4):437–465.

Ryan, J.E.

2003 Race discrimination in education: A legal perspective. *Teachers College Record* 105(6):1087–1118.

Sacerdote, B.

2002 Slavery and the Intergenerational Transmission of Human Capital. Available: http://economics.uchicago.edu/download/slavery.pdf [February 2, 2004].

Sagar, H., and J.W. Schofield

1980 Racial and behavioral cues in black and white children's perceptions of ambiguously aggressive acts. *Journal of Personality and Social Psychology* 39:590–598.

Sampson, R.J., and J.H. Laub

1993 *Crime in the Making: Pathways and Turning Points Through Life*. Cambridge, MA: Harvard University Press.

1997 A life-course theory of cumulative disadvantage and the stability of delinquency. In *Developmental Theories of Crime and Delinquency: Advances in Criminological Theory*, T.P. Thornberry, ed. New Brunswick, NJ: Transaction Publishers.

Sampson, R.J., and J.L. Lauritsen

1997 Racial and ethnic disparities in crime and criminal justice in the United States. *Crime and Justice* 21:311–374.

Sarndal, C., B. Swensson, and J. Wretman
1992 *1997 Model Assisted Survey Sampling.* New York: Springer-Verlag.

Schafer, R.
1979 Racial discrimination in the Boston housing market. *Journal of Urban Economics* 6:176–196.

Schneider, K.T., R.T. Hitlan, and P. Radhakrishnan
2000 An examination of the nature and correlates of ethnic harassment experiences in multiple contexts. *Journal of Applied Psychology* 85(1):3–12.

Schuman, H., E. Singer, R. Donovan, and C. Selltiz
1983 Discriminatory behavior in New York restaurants: 1950 and 1981. *Social Indicators Research* 13:69–83.

Schuman, H., C. Steeh, L. Bobo, and M. Krysan
1997 *Racial Attitudes in America: Trends and Interpretations* (completely revised and updated edition). Cambridge, MA: Harvard University Press.

Sears, D.O., and C.L. Funk
1991 The role of self-interest in social and political attitudes. *Advances in Experimental Social Psychology* 24:1–91.

Sears, D.O., and T. Jessor
1996 Whites' racial policy attitudes: The role of white racism. *Social Science Quarterly* 77:751–759.

Sears, D.O., C. Van Laar, M. Carrillo, and R. Kosterman
1997 Is it really racism? The origins of white Americans' opposition to race-targeted policies. *Public Opinion Quarterly* 61:16–53.

Shadish, W.R., T.D. Cook, and D.T. Campbell
2002 *Experimental and Quasi-Experimental Designs for Generalized Causal Inference.* Boston: Houghton Mifflin.

Shivley, S.
2001 Resurgence of the class action lawsuit in employment discrimination cases: New obstacles presented by the 1991 amendments to the Civil Rights Act. *University of Arkansas, Little Rock, Law Review* 23:925ff.

Sidanius, J., and F. Pratto
1999 *Social Dominance: An Intergroup Theory of Social Hierarchy and Oppression.* New York: Cambridge University Press.

Sigelman, L., and S. Welch
1991 *Black American's Views of Racial Inequality: The Dream Deferred.* New York: Cambridge University Press.

Skrentny, J.D.
1996 *The Ironies of Affirmative Action: Politics, Culture, and Justice in America.* Chicago: University of Chicago Press.

Smelser, N.J., and P.B. Baltes, eds.
2001 *International Encyclopedia of the Social and Behavioral Sciences*, Vol. 7. Oxford, England: Elsevier Science.

Smelser, N.J., W.J. Wilson, and F. Mitchell
2001 Introduction. In National Research Council, *America Becoming: Racial Trends and Their Consequences*, Vol. 1, N.J. Smelser, W.J. Wilson, and F. Mitchell, eds. Commission on Behavioral and Social Sciences and Education. Washington, DC: National Academy Press.

Smith, M.R., and M. Petrocelli
2001 Racial profiling? A multivariate analysis of police traffic stop data. *Police Quarterly* 4:4–27.

Smith, R., and M. Delair
1999 New evidence from lender testing: Discrimination at the pre-application stage. In *Mortgage Lending Discrimination: A Review of Existing Evidence*, M. Turner and F. Skidmore, eds. Urban Institute Monograph Series on Race and Discrimination. Washington, DC: Urban Institute Press.
Smith, T.W.
1993 *Race Relations in New Orleans: An Analysis of a Survey Conducted for the Times Picayune*. Chicago: National Opinion Research Center.
1998 Intergroup relations in contemporary America: An overview of survey research. In *Intergroup Relations in the United States: Research Perspectives*, W. Winborne and R. Cohen, eds. New York: National Conference for Community and Justice.
2000 *Taking America's Pulse: NCCJ's 2000 Survey of Intergroup Relations in the United States*. New York: National Conference for Community and Justice.
2001 *Intergroup Relations in a Diverse America: Data from the 2000 General Social Survey*. New York: American Jewish Committee.
2002 Measuring Racial and Ethnic Discrimination. Unpublished paper, National Opinion Research Center, Chicago.
Smith, W.R., D. Tomaskovic-Devey, M. Mason, M.T. Zingraff, C. Chambers, P. Warren, and C. Wright
2000 "Driving While Black": Establishing a Baseline of Driver Behavior by Measuring Driving Speed and Demographic Characteristics. Unpublished manuscript, North Carolina State University, Raleigh.
Sniderman, P.M.
1985 *Race and Inequality: A Study in American Values*. Chatham, NJ: Chatham House.
Sniderman, P.M., and E.G. Carmines
1997 Reaching beyond race. *Political Science and Politics* 30:466–471.
Sniderman, P.M., and T. Piazza
1993 *The Scar of Race*. Cambridge, MA: Harvard University Press.
Snipp, M.
2000 American Indians and Alaska Natives. In *Encyclopedia of the U.S. Census*, M.J. Anderson, ed. Washington, DC: Congressional Quarterly.
Snyder, M., and W.B. Swann, Jr.
1976 When actions reflect attitudes. The politics of impression management. *Journal of Personality and Social Psychology* 34:1034–1042.
Sobel, M.E.
1995 Causal inference in the social and behavioral sciences. In *Handbook of Statistical Modeling for the Social and Behavioral Sciences*, G. Arminger, C.C. Clogg, and M.E. Sobel, eds. New York: Plenum Press.
1996 An introduction to causal inference. *Sociological Methods and Research* 24:353–379.
Sondik, E.J., J.W. Lucas, J.H. Madans, and S.S. Smith
2000 Race/ethnicity and the 2000 census: Implications for public health. *American Journal of Public Health* 90(11):1709–1713.
Spirtes, P., C. Glymour, and R. Scheines
1993 *Causation, Prediction, and Search*. (Spring Lecture Notes in Statistics, No. 81.) New York: Springer-Verlag.
Spitzer, E.
1999 The New York City Police Department's Stop & Frisk Practices: A Report to the People of the State of New York from the Office of the Attorney General. Available: http://www.oag.state.ny.us/press/reports/stop_frisk/stop_frisk.html [January 29, 2004].

Squires, G.D.
1994 *Capital and Communities in Black and White*. Albany: SUNY Press.
Squires, G.D., and S. O'Connor
2001 *Color and Money*. Albany: SUNY Press.
Squires, G., and W. Velez
1988 Insurance redlining and the process of discrimination. *Review of Black Political Economy* 16:63–75.
St. Jean, Y., and J.R. Feagin
1998 The family cost of white racism: The case of African American families. *Journal of Comparative Family Studies* 29:297–312.
1999 *Double Burden: Black Women and Everyday Racism*. Armonk, NY: M.E. Sharpe.
Stackhouse, H.F, and S. Brady
2003 *Census 2000 Mail Response Rates*. (Census 2000 Evaluation A.7.a.) Washington, DC: U.S. Census Bureau.
Staub, E.
1989 *The Roots of Evil: The Origins of Genocide and Other Group Violence*. Cambridge: Cambridge University Press.
Steeh, C., and M. Krysan
1996 Affirmative action and the public, 1970–1985. *Public Opinion Quarterly* 60:128–158.
Stephan, W.G.
1985 Intergroup relations. In *The Handbook of Social Psychology*, 3rd ed., Vol. II, G. Lindzey and E. Aronson, eds. New York: Random House.
Stephan, W.G., and D. Rosenfield
1982 Racial and ethnic stereotypes. In *In the Eye of the Beholder*, A.G. Miller, ed. New York: Praeger.
Stephan, W.G., and C.W. Stephan
1989 Emotional reactions to interracial achievement outcomes. *Journal of Applied Social Psychology* 19:608–621.
Strom, K.J.
2001 *Hate Crime Reported in NIBRS, 1997–1999*. Washington, DC: U.S. Bureau of Justice Statistics.
Sudman, S., and N.M. Bradburn
1982 *Asking Questions*. San Francisco, CA: Jossey-Bass.
Sue, S., and S. Okazaki
1990 Asian American educational achievements: A phenomenon in search of an explanation. *American Psychologist* 45(8):913–920.
Sue, S., D.W. Sue, and D. Sue
1975 Asian Americans as a minority group. *American Psychologist* 30:906–910.
Suh, S.A.
2000 Women's perceptions of workplace discrimination: Impacts of racial group, gender, and class. In *Prismatic Metropolis: Inequality in Los Angeles*, L.D. Bobo, M.L. Oliver, J.H. Johnson, Jr., and A. Valenzuela, Jr., eds. New York: Russell Sage Foundation.
Supphellen, M., O.A. Kvitastein, and S.T. Johansen
1997 Projective questioning and ethnic discrimination: A procedure for measuring employer bias. *Public Opinion Quarterly* 61:208–224.
Talaska, C.A., S.T. Fiske, and S. Chaiken
2003 Stereotypes, Emotional Prejudices, and the Prediction of Discriminatory Behavior. Unpublished meta-analysis, Princeton University, Princeton, NJ.

Taylor, D.M., S.C. Wright, and K. Ruggiero
1991 The personal/group discrimination discrepancy: Responses to experimentally induced personal and group discrimination. *Journal of Social Psychology* 131:847–858.

Texas Department of Public Safety
2000 Traffic Stop Data Report. Available: http://www.txdps.state.tx.us/director_staff/public_information/trafrep2q00.pdf [January 29, 2004].

Thompson, M.S., and V.M. Keith
2001 The blacker the berry: Gender, skin tone, self-esteem, and self-efficacy. *Gender and Society* 15(3):336–357.

Thornton, R.
1987 *American Indian Holocaust and Survival: A Population History Since 1492.* Norman: University of Oklahoma Press.
2001 Trends among American Indians in the United States. In *America Becoming: Racial Trends and Their Consequences*, Vol. 1, N.J. Smelser, W.J. Wilson, and F. Mitchell, eds. Commission on Behavioral and Social Sciences and Education. Washington, DC: National Academy Press.

Tonry, M.
1995 *Malign Neglect: Race, Crime, and Punishment in America.* New York: Oxford University Press.
1996 *Sentencing Matters.* New York: Oxford University Press.

Toplin, R.B.
1974 *Slavery and Race Relations in Latin America.* Westport, CT: Greenwood Press.

Tucker, C., and R. Harrison
1995 The research agenda on issues surrounding the definition of racial and ethnic categories. In *Proceedings of the Section on Government Statistics.* Washington, DC: American Statistical Association.

Tucker, C., and B. Kojetin
1996 Testing racial and ethnic origin questions in the CPS supplement. *Monthly Labor Review* (September):3–7.

Tucker, C., S. Miller, and J. Parker
2000 Comparing census race data under the old and the new standards. Unpublished paper, Bureau of Labor Statistics, Washington, DC.

Turner, M.A., and F. Skidmore
1999 *Mortgage Lending Discrimination: A Review of Existing Evidence.* (Urban Institute Monograph Series on Race and Discrimination.) Washington, DC: Urban Institute Press.

Turner, M.A., R. Struyk, and J. Yinger
1991a *Housing Discrimination in America: Summary of Findings from the Housing Discrimination Study.* Washington, DC: Urban Institute Press.

Turner, M.A., M. Fix, and R.J. Struyk
1991b *Opportunities Denied, Opportunities Diminished: Racial Discrimination in Hiring.* Washington, DC: Urban Institute Press.

Turner, M.A., F. Freiberg, E. Godfrey, C. Herbig, D.K. Levy, and R.R. Smith
2002a *All Other Things Being Equal: A Paired Testing Study of Mortgage Lending Institution.* Washington, DC: U.S. Department of Housing and Urban Development.

Turner, M.A., S.L. Ross, G.C. Galster, and J. Yinger
2002b *Discrimination in Metropolitan Housing Markets: National Results from Phase I of HDS2000.* Washington, DC: U.S. Department of Housing and Urban Development.

Turner, M.A., B.A. Bednarz, C. Herbig, and S.J. Lee

2003 *Discrimination in Metropolitan Housing Markets: Phase 2, Asians and Pacific Islanders.* Washington, DC: Urban Institute Press.

Tyler, J.H., R.J. Murnane, and J.B. Willett

1998 Estimating the Impact of the GED on the Earnings of Young Dropouts Using a Series of Natural Experiments. NBER Working Paper 6391. Available: http://www.nber.org/papers/w6391 [January 29, 2004].

U.S. Census Bureau

1993 We the Americans: Asians. Available: http://www.census.gov/apsd/wepeople/we-3.pdf [January 29, 2004].

2001a Population by Race and Hispanic or Latino Origin, for All Ages and for 18 Years and Over, for the United States: 2000. Available: http://www.census.gov/Press-Release/www/2001/tables/st00_1.pdf [January 29, 2004].

2001b Overview of Race and Hispanic Origin. Census 2000 Brief. Available: http://www.census.gov/prod/2001pubs/c2kbr01-1.pdf [January 29, 2004].

2002 Statistical Abstract of the United States, 2001. Available: http://www.census.gov/prod/www/statistical-abstract-02.html [January 29, 2004].

U.S. Department of Education

2001a *The Condition of Education, 2001.* (NCES 2001-125.) Washington, DC: U.S. Government Printing Office.

2001b *Students Whose Parents Did Not Go to College: Postsecondary Access, Persistence, and Attainment.* (NCES 2001-126 by S. Choy.) Washington, DC: National Center for Education Statistics. Available: http://nces.ed.gov/pubs2001/2001126.pdf [January 29, 2004].

U.S. Department of Justice

2000 Crime in the United States, 2000. Uniform Crime Reports. Available: http://www.fbi.gov/ucr/cius_00/contents.pdf [January 29, 2004].

2001 Homicide Trends in the U.S.: Trends by Race. Available: http://www.ojp.usdoj.gov/bjs/homicide/race.htm [January 29, 2004].

U.S. Equal Employment Opportunity Commission

2002 Charge Statistics, FY 1992 through FY 2002. Available: www.eeoc.gov/stats/charges.html [January 29, 2004].

U.S. Office of Management and Budget

1977 *Statistical Policy Directive No. 15, Race and Ethnic Standards for Federal Statistics and Administrative Reporting.* Washington, DC: U.S. Office of Management and Budget.

1997 *Revisions to the Standard for the Classification of Federal Data on Race and Ethnicity.* Washington, DC: U.S. Office of Management and Budget.

2000 Guidance on Aggregation and Allocation of Data on Race, OMB Bulletin No. 00-02. Available: http://www.whitehouse.gov/omb/bulletins/b00-02.html [January 29, 2004].

van den Berghe, P.

1967 *Race and Racism: A Comparative Perspective.* New York: John Wiley & Sons.

Van Hook, J.

2002 Immigration and African American educational opportunity: Are limited English proficient students disproportionately transforming minority schools? *Sociology of Education* 75:169–189.

Verniero, P., and P.H. Zoubek

1999 *Interim Report of the State Police Review Team Regarding Allegations of Racial Profiling.* Trenton, NJ: Attorney General, State of New Jersey.

Waldinger, R., and M.I. Lichter
2003 *How the Other Half Works: Immigration and the Social Organization of Labor.* Berkeley: University of California Press.
Walker, S.
2001 Searching for the denominator: Problems with police traffic stop data and an early warning system solution. *Justice Research and Policy* 3:63–95.
Walker, S., C. Spohn, and M. DeLone
1996 *The Color of Justice: Race, Ethnicity, and Crime in America.* Belmont, CA: Wadsworth.
Washington Post
2002 Lawsuits accuse 4 airlines of bias: Men say perceived ethnicity got them taken off flights. June 5, p. A01.
Washington State Patrol
2001 *Report to the Legislature on Routine Traffic Stop Data.* Olympia: Washington State Patrol and Criminal Justice Training Commission.
Webb, G.
1999 Driving while black. *Esquire* (April):118–127.
Weich, R., and C. Angulo
2002 Racial disparities in the American criminal justice system. In *Rights at Risk: Equality in an Age of Terrorism*, D.M. Piche, W.L. Taylor, and R.A. Reed, eds. Washington, DC: Citizen's Civil Rights Commission.
Weinberger, C., and L. Joy
2003 Wage gaps among black college graduates, 1980–2001. Unpublished paper, University of California, Santa Barbara.
Weisberg, S.
1985 *Applied Linear Regression* 2nd ed. New York: John Wiley & Sons.
Weiss, Y., and R. Gronau
1981 Expected interruptions in labour force participation and sex-related differences in earnings growth. *Review of Economic Studies* 48(4):607–619.
Weitzer, R., and S.A. Tuch
1999 Race, class, and perceptions of discrimination by the police. *Crime and Delinquency* 45:494–507.
Welner, K.G.
2001 *Legal Rights, Local Wrongs. When Community Control Collides with Educational Equity.* Albany: SUNY Press.
Western, B.
2002 The impact of incarceration on wage mobility and inequality. *American Sociological Review* 67:477–498.
Western, B., and B. Pettit
2002 Beyond crime and punishment: Prisons and inequality. *Contexts* 1(3):37–43.
White, B.
2000 ACLU suit accuses U.S. Customs of "profiling." *Washington Post*, May 14.
Wienk, R., C.E. Reid, J.C. Simonson, and F.J. Eggers
1979 *Measuring Racial Discrimination in American Housing Markets: The Housing Practices Survey.* Washington, DC: U.S. Department of Housing and Urban Development.
Williams, D.R., and C.A. Collins
1995 U.S. socioeconomic and racial differences in health. *Annual Review of Sociology* 21:349–386.

Williams, D.R., and H.W. Neighbors
2001 Racism, discrimination, and hypertension: Evidence and needed research. *Ethnicity and Disease* 11(Supplement):800–816.

Williams, D.R., and R. Williams-Morris
2000 Racism and mental health: The African American experience. *Ethnicity and Health* 5(3/4):243–268.

Williams, D.R., H.W. Neighbors, and J.S. Jackson
2003 Racial/ethnic discrimination and health: Findings from community studies. *American Journal of Public Health* 93(2):200–208.

Wilson, W.J.
1987 *The Truly Disadvantaged.* Chicago: University of Chicago Press.

Wilson, J.Q., and H. Higgins
2002 It isn't easy being screened. *The Wall Street Journal*, January 10.

Winant, H.
2001 *The World Is a Ghetto: Race and Democracy Since World War II.* New York: Basic Books.

Winship, C., and S.L. Morgan
1999 The estimation of causal effects from observational data. *Annual Review of Sociology* 25:659–707.

Winship, C., and M. Sobel
2004 Causal analysis in sociological studies. In *Handbook of Data Analysis*, M. Hardy and A. Bryman eds. Thousand Oaks, CA: Sage.

Wissoker, D.A., W. Zimmerman, and G. Galster
1997 *Testing for Discrimination in Home Insurance.* Washington, DC: Urban Institute Press.

Wittenbrink, B., C.M. Judd, and B. Park
1997 Evidence for racial prejudice at the implicit level and its relationship with questionnaire measures. *Journal of Personality and Social Psychology* 72:262–274.
2001 Evaluative versus conceptual judgments in automatic stereotyping and prejudice. *Journal of Personality and Social Psychology* 37:244–252.

Wood, J.
1994 Is "symbolic racism" racism? A review informed by intergroup behavior. *Journal of Personality and Social Psychology* 72:262–274.

Word, C.O., M.P. Zanna, and J. Cooper
1974 The nonverbal mediation of self-fulfilling prophecies in interracial interaction. *Journal of Experimental Social Psychology* 10:109–120.

Wyly, E.K., and S.R. Holloway
1999 The new color of money: Neighborhood lending patterns in Atlanta revisited. *Housing Facts and Findings* 1(2):1–11.

Yen, I.H., D.R. Ragland, B.A. Greiner, and J.M. Fisher
1999 Workplace discrimination and alcohol consumption: Findings from the San Francisco Muni Health and Safety Study. *Ethnicity and Disease* 9:70–80.

Yinger, J.
1986 Measuring racial discrimination with fair housing audits: Caught in the act. *American Economic Review* 76:881–893.
1993 Access denied, access constrained: Results and implications of the 1989 housing discrimination study. In *Clear and Convincing Evidence*, M. Fix and R.J. Struyk, eds. Washington, DC: Urban Institute Press.
1995 *Closed Doors, Opportunities Lost: The Continuing Costs of Housing Discrimination.* New York: Russell Sage Foundation.

Zax, J.
1989 Race and commutes. *Journal of Urban Economics* 28:336–348.
Zingraff, M.T., H.M. Mason, W.R. Smith, D. Tomaskovic-Devey, P. Warren, H.L. McMurray, and C.R. Fenlon
2000 Evaluating North Carolina State Highway Patrol Data: Citations, Warnings, and Searches in 1998. Available: http://www.nccrimecontrol.org/shp/ncshpreport.htm [January 29, 2004].
Zuckerman, M.
1990 Some dubious premises in research and theory on racial differences. *American Psychologist* 45:1297–1303.
Zuberi, T.
2001 *Thicker Than Blood: How Racial Statistics Lie.* Minneapolis: University of Minnesota Press.
Zweigenhaft, R.L., and G.W. Domhoff
1991 *Blacks in the White Establishment: A Study of Race and Class in America.* New Haven, CT: Yale University Press.

Selected Bibliography

Below we provide a list of selected papers from the theoretical and empirical literature on discrimination. These papers are organized by the specific domains discussed in Chapter 4.

GENERAL

Ayres, I., and P. Siegelman
1995 Race and gender discrimination in bargaining for a new car. *American Economic Review* 85(3):304–321.

Becker, G.S.
1971 *The Economics of Discrimination*, 2nd ed. Chicago: University of Chicago Press.

Brown, T.N.
2001 Measuring self-perceived racial and ethnic discrimination in a social survey. *Sociological Spectrum* 21:377–392.

Coate, S., and G.C. Loury
1993 Will affirmative-action policies eliminate negative stereotypes? *American Economic Review* 83(5):1220–1240.

Crosby, F., S. Bromley, and L. Saxe
1980 Recent unobtrusive studies of black and white discrimination and prejudice: A literature review. *Psychological Bulletin* 87:546–563.

Dovidio, J.F., and S.L. Gaertner, eds.
1986 Prejudice, discrimination, and racism: Historical trends and contemporary approaches. In *Prejudice, Discrimination, and Racism*. San Diego, CA: Academic Press.

Dovidio, J.F., J.C. Brigham, B.T. Johnson, and S.L. Gaertner
1996 Stereotyping, prejudice, and discrimination: Another look. In *Stereotypes and Stereotyping*, C.N. Macrae, C. Stangor, and M. Hewstone, eds. New York: Guilford Press.

Feagin, J.R.
1991 The continuing significance of race: Antiblack discrimination in public places. *American Sociological Review* 56:101–116.
Fiske, S.T.
1998 Stereotyping, prejudice, and discrimination. In *The Handbook of Social Psychology*, 4th ed., D.T. Gilbert, S.T. Fiske, and G. Lindzey, eds. New York: McGraw-Hill.
2000 Stereotyping, prejudice, and discrimination at the seam between centuries: Evolution, culture, mind, and brain. *European Journal of Social Psychology* 30:299–322.
Fix, M., and R.J. Struyk
1993 *Clear and Convincing Evidence: Measurement of Discrimination in America.* Washington, DC: Urban Institute Press.
Fix, M., and M.A. Turner
1998 *A National Report Card on Discrimination in America: The Role of Testing.* Washington, DC: Urban Institute Press.
National Research Council
1989 *A Common Destiny: Blacks and American Society*, G.D. Jaynes and R.N. Williams, Jr., eds. Committee on the Status of Black Americans, Commission on Behavioral and Social Sciences and Education. Washington, DC: National Academy Press.
2001 *America Becoming: Racial Trends and Their Consequences*, Vol. II, N.J. Smelser, W.J. Wilson, and F. Mitchell, eds. Commission on Behavioral and Social Sciences and Education. Washington, DC: National Academy Press.
Phelps, E.S.
1972 The statistical theory of racism and sexism. *American Economic Review* 62:659–661.

CRIMINAL JUSTICE

Ayres, I., and J. Waldfogel
1994 A market test for race discrimination in bail setting. *Stanford Law Review* 46:987–1047.
Blumstein, A.
1982 On the racial disproportionality of United States' prison populations. *Journal of Criminal Law and Criminology* 73:1259–1281.
1993 Racial disproportionality of U.S. prison populations revisited. *University of Colorado Law Review* 64:743–760.
Bright, S.
1995 Discrimination, death, and denial: The tolerance of racial discrimination in infliction of the death penalty. *Santa Clara Law Review* 35(2):433–483.
Knowles, J., and N. Persico
2001 Racial bias in motor vehicle searches: Theory and evidence. *Journal of Political Economy* 109:203–229.
Knowles, J., N. Persico, and P. Todd
1999 *Racial Bias in Motor Vehicle Searches: Theory and Evidence.* (NBER Working Paper 7449.) Cambridge, MA: National Bureau of Economic Research.
National Research Council
2001 Racial trends in the administration of justice. In *America Becoming: Racial Trends and Their Consequences*, Vol. II, N.J. Smelser, W.J. Wilson, and F. Mitchell, eds. Commission on Behavioral and Social Sciences and Education. Washington, DC: National Academy Press.

Sampson, R.J., and J.L. Lauritsen
1997 Racial and ethnic disparities in crime and criminal justice in the United States. *Crime and Justice* 21:311–374.
Tonry, M.
1995 *Malign Neglect: Race, Crime, and Punishment in America.* New York: Oxford University Press.
Weich, R., and C. Angulo
2002 Racial disparities in the American criminal justice system. In *Rights at Risk: Equality in an Age of Terrorism*, D.M. Piche, L. Taylor, and R.A. Reed, eds. Washington, DC: Citizen's Civil Rights Commission.
Weitzer, R.
1996 Racial discrimination in the criminal justice system: Findings and problems in the literature. *Journal of Criminal Justice* 24(4):309–322.

EDUCATION

Hallinan, M.T.
2001 Sociological perspectives on black-white inequalities in American schooling. *Sociology of Education* Special Issue:50–70.
National Research Council
2002 *Minority Students in Special and Gifted Education*, M.S Donovan and C.T. Cross, eds. Committee on Minority Representation in Special Education, Division of Behavioral and Social Sciences and Education. Washington, DC: National Academy Press.
Norman, O., C.R. Ault, B. Bentz, and L. Meskimen
2001 The black-white "achievement gap" as a perennial challenge of urban science education: A sociocultural and historical overview with implications for research and practice. *Journal of Research in Science Teaching* 38:1101–1114.
Oswald, D.P., M.J. Coutinho, and A.M. Best
2000 Community and Social Predictors of Overrepresentation of Minority Children in Special Education. Unpublished paper presented at the Harvard Civil Rights Project Conference on Minority Issues in Special Education, Cambridge, MA, November 17.
Selmi, M.
1999 The shape of the river: Long-term consequences of considering race in college and university admissions. *Virginia Law Review* 85:697–739.

HEALTH CARE

American Medical Association Council on Ethical and Judicial Affairs
1990 Black-white disparities in health care. *Journal of the American Medical Association* 263(17):2344–2346.
Ford, E.S., and R.S. Cooper
1995 Racial/ethnic differences in health care utilization of cardiovascular procedures: A review of the evidence. *Health Services Research* 30:237–252.
Gomes, C., and T.G. McGuire
2001 Identifying the Sources of Racial and Ethnic Differences in Health Care Use. Unpublished paper, Boston University, Boston, MA.
Institute of Medicine
2003 *Unequal Treatment: Confronting Racial and Ethnic Disparities in Health Care*, B.D.

Smedley, A.Y. Smith, and A.R. Nelson, eds. Committee on Understanding and Eliminating Racial and Ethnic Disparities in Health Care, Board on Health Sciences Policy. Washington, DC: The National Academies Press.

Kressin, N.R., and L.A. Petersen
2001 Racial differences in the use of invasive cardiovascular procedures: Review of the literature and prescription for further research. *Annals of Internal Medicine* 135(5):352–366.

Krieger, N.
1999 Embodying inequality: A review of concepts, measures and methods for studying health consequences of discrimination. *International Journal of Health Services* 29:295–352.
2000 Discrimination and health. In *Social Epidemiology*, L.F. Berkman and I. Kawachi, eds. New York: Oxford University Press.

Landrine, H., and E.A. Klonoff
1996 The schedule of racist events: A measure of racial discrimination and a study of its negative physical and mental health consequences. *Journal of Black Psychology* 22:144–168.

LaVeist, T., and M.C. Gibbons
2001 Measuring racial and ethnic discrimination in the U.S. healthcare setting: A review of the literature and suggestions for a monitoring program. Final report to the U.S. Department of Health and Human Services, April 12.

Lillie-Blanton, M., P.E. Parson, H. Gayle, and A. Dievler
1996 Racial differences in health: Not just black and white, but shades of gray. *Annual Review of Public Health* 17:411–448.

Lillie-Blanton, M., M. Brodie, D. Rowland, D. Altman, and M. McIntosh
2000 Race, ethnicity, and the health care system: Public perceptions and experiences. *Medical Care Research and Review* 57(1):218–235.

Mayberry, R., F. Mili, and E. Ofili
2000 Racial and ethnic differences in access to medical care. *Medical Care Research and Review, 2000, Supplement I* 57:108–146.

Sheifer, E.S., J.J. Escarce, and K.A. Schulman
2000 Race and sex differences in the management of coronary artery disease. *American Heart Journal* 139(5):848–857.

Williams, D.R.
1999 Race, socioeconomic status, and health: The added effects of racism and discrimination. *Annals of the New York Academy of Sciences* 896:173–188.

Williams, D.R., and T.D. Rucker
2000 Understanding and addressing racial disparities in health care. *Health Care Financing Review* 21:75–90.

HOUSING/MORTGAGE LENDING

Carr, J.H., and I.F. Megbolugbe
1993 Federal Reserve Bank of Boston study on mortgage lending revisited. *Journal of Housing Research* 4:277–314.

Center for Community Change
1989 Mortgage Lending Discrimination Testing Project. Unpublished report, U.S. Department of Housing and Urban Development, Washington, DC.

Charles, K., and E. Hurst
2002 The transition to home ownership and the black-white wealth gap. *Review of Economics and Statistics* 84(2):281–297.

Donnerstein, E., M. Donnerstein, and C. Koch
1975 Racial discrimination in apartment rentals: A replication. *Journal of Social Psychology* 96:37–38.
Ferguson, M.F., and S.R. Peters
1995 What constitutes evidence of discrimination in lending? *Journal of Finance* 50:739–748.
Fix, M., G. Galster, and R.J. Struyk
1993 An overview of auditing for discrimination. In *Clear and Convincing Evidence: Measurement of Discrimination in America*, M. Fix and R.J. Struyk, eds. Washington, DC: Urban Institute Press.
Galster, G.
1990a Federal fair housing policy: The great misapprehension. In *Building Foundations: Housing and Federal Policy*, D. Dipasquale and L.C. Keyes, eds. Philadelphia: University of Pennsylvania Press.
1990b Racial steering by real estate agents: A review of the audit evidence. *Review of Black Political Economy* 18:105–129.
1992 Research on discrimination in housing and mortgage markets: Assessment and future directions. *Housing Policy Debate* 3:639–684.
Heckman, J.J., and P. Siegelman
1993 The Urban Institute audit studies: Their methods and findings. In *Clear and Convincing Evidence: Measurement of Discrimination in America*, M. Fix and R.J. Struyk, eds. Washington, DC: Urban Institute Press.
Ladd, H.F.
1998 Evidence on discrimination in mortgage lending. *Journal of Economic Perspectives* 12:41–62.
Massey, D.S., and N.A. Denton
1993 *American Apartheid: Segregation and the Making of the Underclass.* Cambridge, MA: Harvard University Press.
Munnell, A.H., G.M.B. Tootell, L.E. Browne, and J. McEneaney
1996 Mortgage lending in Boston: Interpreting HMDA data. *American Economic Review* 86(1):25–53.
National Research Council
2001 Residential segregation and neighborhood conditions in U.S. metropolitan areas. In *America Becoming: Racial Trends and Their Consequences*, Vol. 1, N.J. Smelser, W.J. Wilson, and F. Mitchell, eds. Commission on Behavioral and Social Sciences and Education. Washington, DC: National Academy Press.
2002 *Measuring Housing Discrimination in a National Study: Report of a Workshop*, A.W. Foster, F. Mitchell, and S.E. Fienberg, eds. Committee on National Statistics, Division of Behavioral and Social Sciences and Education. Washington, DC: National Academy Press.
Nesiba, R.F.
1996 Racial discrimination in residential lending markets: Why empirical researchers always see it and economic theorists never do. *Journal of Economic Issues* 30(1):51–78.
Ondrich, J., S.L. Ross, and J. Yinger
2000 How common is housing discrimination? Improving on traditional measures. *Journal of Urban Economics* 47:470–500.
2001 Geography of housing discrimination. *Journal of Housing Research* 12(2):217–238.
Ross, S.L.
2002 Paired testing and the 2000 Housing Discrimination Survey. In *Measuring Housing Discrimination in a National Study: Report of a Workshop*, A.W. Foster, F. Mitchell,

and S.E. Fienberg, eds. National Research Council. Washington, DC: National Academy Press.

Ross, S.L., and J. Yinger

1999 Does discrimination exist? The Boston Fed study and its critics. In *Mortgage Lending Discrimination: A Review of Existing Evidence*, M. Turner and F. Skidmore, eds. (Urban Institute Monograph Series on Race and Discrimination.) Washington, DC: Urban Institute Press.

2002 *The Color of Credit: Mortgage Discrimination, Research Methodology, and Fair-Lending Enforcement.* Cambridge, MA: MIT Press.

Schafer, R.

1979 Racial discrimination in the Boston housing market. *Journal of Urban Economics* 6:176–196.

Smith, R., and M. Delair

1999 New evidence from lender testing: Discrimination at the pre-application stage. In *Mortgage Lending Discrimination: A Review of Existing Evidence*, M. Turner and F. Skidmore, eds. (Urban Institute Monograph Series on Race and Discrimination.) Washington, DC: Urban Institute Press.

Squires, G., and W. Velez

1988 Insurance redlining and the process of discrimination. *Review of Black Political Economy* 16:63–75.

Turner, M.A., and F. Skidmore

1999 *Mortgage Lending Discrimination: A Review of Existing Evidence.* (Urban Institute Monograph Series on Race and Discrimination.) Washington, DC: Urban Institute Press.

Turner, M.A., R.J. Struyk, and J. Yinger

1991 *Housing Discrimination in America: Summary of Findings from the Housing Discrimination Study*. Washington, DC: Urban Institute Press.

Turner, M.A., F. Freiberg, E. Godfrey, C. Herbig, D.K. Levy, and R.R. Smith

2002 *All Other Things Being Equal: A Paired Testing Study of Mortgage Lending Institutions.* Washington, DC: U.S. Department of Housing and Urban Development.

Turner, M.A., S.L. Ross, G.C. Galster, and J. Yinger

2002 *Discrimination in Metropolitan Housing Markets: National Results from Phase I of HDS2000.* Washington, DC: U.S. Department of Housing and Urban Development.

Wienk, R., C.E. Reid, J.C. Simonson, and F.J. Eggers

1979 *Measuring Racial Discrimination in American Housing Markets: The Housing Practices Survey.* Washington, DC: U.S. Department of Housing and Urban Development.

Wyly, E.K., and S.R. Holloway

1999 The new color of money: Neighborhood lending patterns in Atlanta revisited. *Housing Facts and Findings* 1(2):1–11.

Yinger, J.

1986 Measuring racial discrimination with fair housing audits: Caught in the act. *American Economic Review* 76:881–893.

1995 *Closed Doors, Opportunities Lost: The Continuing Costs of Housing Discrimination.* New York: Russell Sage Foundation.

1996 Discrimination in mortgage lending: A literature review. In *Mortgage Lending, Racial Discrimination, and Federal Policy*, J. Goering and R. Wienk, eds. Washington, DC: Urban Institute Press.

LABOR MARKETS

Altonji, J.G., and R. Blank
1999 Race and gender in the labor market. *Handbook of Labor Economics* 3:3143–3259.
Altonji, J.G., and U. Doraszelski
2002 The Role of Permanent Income and Demographics in Black/White Differences in Wealth. Yale University Economic Growth Center Discussion Paper No. 850. Available: http://www.library.yale.edu/socsci/egcdp850.txt [January 29, 2004].
Arrow, K.
1973 The theory of discrimination. In *Discrimination in Labor Markets*, O.A. Ashenfelter and A. Rees, eds. Princeton, NJ: Princeton University Press.
Baldwin, M.L., and W.G. Johnson
1996 The employment effects of wage discrimination against black men. *Industrial and Labor Relations Review* 49(January):302–316.
Becker, G.S.
1971 *The Economics of Discrimination*, 2nd ed. Chicago: University of Chicago Press.
Bendick, M., Jr., C. Jackson, and V. Reinoso
1994 Measuring employment discrimination through controlled experiments. *Review of Black Political Economy* 23:25–48.
Black, D.A.
1995 Discrimination in an equilibrium search model. *Journal of Labor Economics* 13(2):309–334.
Bobo, L.D., and S.A. Suh
2000 Surveying racial discrimination: Analyses from a multiethnic labor market. In *Prismatic Metropolis: Inequality in Los Angeles*, L.D. Bobo, M.L. Oliver, J.H. Johnson, Jr., and A. Valenzuela, Jr., eds. New York: Russell Sage Foundation.
Borjas, G.J., and S.G. Bronars
1989 Consumer discrimination and self-employment. *Journal of Political Economy* 97(3):581–606.
Bowlus, A.J., and Z. Eckstein
2002 Discrimination and skill differences in an equilibrium search model. *International Economic Review* 43(4):1309–1345.
Braddock, J.H., and J.M. McPartland
1987 How minorities continue to be excluded from equal employment opportunities: Research on labor market and institutional barriers. *Journal of Social Issues* 43:5–39.
Cain, G.G.
1986 The economic analysis of labor market discrimination: A survey. In *Handbook of Labor Economics*, Vol. 1, O. Asherfelter and R. Layard, eds. Amsterdam: Elsevier/North-Holland.
Cross, H., G. Kenney, J. Mell, and W. Zimmerman
1990 *Employer Hiring Practices: Differential Treatment of Hispanic and Anglo Job Seekers*. Washington, DC: Urban Institute Press.
Darity, W.A., Jr., and P.L. Mason
1998 Evidence on discrimination in employment: Codes of color, codes of gender. *Journal of Economic Perspectives* 12:63–90.
Dempster, A.
1988 Employment discrimination and statistical science. *Statistical Science* 3:149–161.
Donohue, J.J., and J. Heckman
1991 Continuous versus episodic change: The impact of civil rights policy on the economic status of blacks. *Journal of Economic Literature* 29:1603–1643.

Heckman, J.J.

1998 Detecting discrimination. *Journal of Economic Perspectives* 12:101–116.

Heckman, J.J., and B.S. Payner

1989 Determining the impact of federal antidiscrimination policy on the economic status of blacks: A study of South Carolina. *American Economic Review* 79(1):138–177.

Heckman, J.J., and P. Siegelman

1993 The Urban Institute audit studies: Their methods and findings. In *Clear and Convincing Evidence: Measurement of Discrimination in America*, M. Fix and R.J. Struyk, eds. Washington, DC: Urban Institute Press.

Kirschenman, J., and K.M. Neckerman

1991 "We'd love to hire them, but . . ." The meaning of race for employers. In *The Urban Underclass*, C. Jencks and P.E. Peterson, eds. Washington, DC: Brookings Institution.

Lundberg, S.J., and R. Startz

1983 Private discrimination and social intervention in competitive labor markets. *American Economic Review* 73:340–347.

Oliver, M., and T. Shapiro

1995 *Black Wealth, White Wealth: A New Perspective on Racial Inequality*. New York: Routledge.

Shulman, S., and W. Darity, Jr.

1989 *The Question of Discrimination: Racial Inequality in the U.S. Labor Market*. Middletown, CN: Wesleyan University Press.

Turner, M.A., M. Fix, and R.J. Struyk

1991 *Opportunities Denied and Opportunities Diminished: Racial Discrimination in Hiring*. Washington, DC: Urban Institute Press.

Wilson, W.J.

1987 *The Truly Disadvantaged*. Chicago: University of Chicago Press.

Appendix A

Workshop Agenda

NATIONAL RESEARCH COUNCIL

COMMITTEE ON NATIONAL STATISTICS

CENTER FOR EDUCATION

Workshop on Measuring Racial Disparities and Discrimination in Elementary and Secondary Education

Georgetown Facility
Cecil and Ida Green Building
Room 104
2001 Wisconsin Avenue, NW
Washington, DC

July 1, 2002

8:30 Breakfast in Meeting Room

9:00 Welcome and Introductions

Samuel Lucas*, Member, Panel on Methods for Assessing Discrimination*
Andy White*, Director, Committee on National Statistics*
Jeanette Lim*, Director, Program Legal Group, Office for Civil Rights, U.S. Department of Education*

9:30 **Session One: What Constitutes Race Discrimination in Education? A Social Science Perspective**
- Presenter: **Roslyn A. Mickelson** (University of North Carolina-Charlotte)
- Discussant: **Valerie Lee** (University of Michigan)

10:00 General Discussion: Q & A

10:45 **Session Two: What Constitutes Race Discrimination in Education? A Legal Perspective**
- Presenter: **James Ryan** (University of Virginia Law School)
- Discussant: **Michael Rebell** (Campaign for Fiscal Equity, Inc.)

11:15 General Discussion: Q & A

12:00 Lunch

1:00 **Session Three: Racial Disparities and Discrimination in Education: What Do We Know, How Do We Know It, and What Do We Need to Know?**
- Presenter: **George Farkas** (Pennsylvania State University)
- Discussant: **Ronald Ferguson** (Harvard University)

1:30 General Discussion: Q & A

2:15 **Session Four: Measuring Discrimination: Alternative Techniques and Applications from Other Domains**
- Presenters: **Harry Holzer** (Georgetown University) and **Jens Ludwig** (Georgetown University)
- Discussant: **Judith Hellerstein** (University of Maryland)

2:45 General Discussion: Q & A

3:30 **Session Five: Applications and Directions for the Department of Education**
Panelists: **Joan First** (National Coalition of Advocates for Students)
Willis Hawley (University of Maryland)
John Kain (University of Texas-Dallas)

Gerald Reynolds* (Assistant Secretary, Office for Civil Rights, U.S. Department of Education)
Marilyn McMillen Seastrom (Chief Statistician, National Center for Education Statistics, U.S. Department of Education)

4:15 Closing Remarks
Rebecca Blank*, Chair, Panel on Methods for Assessing Discrimination*

4:30 Adjourn

*Could not attend; Dan Sutherland, Chief of Staff, Office for Civil Rights, U.S. Department of Education, came in his place.

Appendix B

Biographical Sketches

REBECCA M. BLANK (*Chair*) is dean of the Gerald R. Ford School of Public Policy, the Henry Carter Adams Collegiate Professor of Public Policy, and professor of economics. Her research focuses on the interaction among the macroeconomy, government antipoverty programs, and the behavior and well-being of low-income families. Her publications include *Social Protection vs. Economic Flexibility: Is There a Trade Off?*, which compares the social protection programs in the United States and other industrialized countries, and *It Takes a Nation: A New Agenda for Fighting Poverty*, which analyzes recent discussion about poverty and public policy in the United States. Professor Blank joined the Ford School faculty after serving as a member of the President's Council of Economic Advisors in Washington, D.C. A graduate of the University of Minnesota, Professor Blank received a Ph.D. from the Massachusetts Institute of Technology.

JOSEPH G. ALTONJI is the Thomas DeWitt Cuyler Professor of Economics at Yale University. He has held previous faculty positions at Columbia University and Northwestern University. Professor Altonji is also a research associate at the National Bureau of Economic Research. He is a fellow of the Econometric Society, served on the board of editors of the American Economic Review and as coeditor of the *Journal of Human Resources*, and is currently an associate editor of *Econometrica.* He received B.A. and M.A. degrees in economics from Yale University and a Ph.D. in economics from Princeton University. Professor Altonji specializes in labor economics and applied econometrics. In recent years, he has focused on the role of family background, school characteristics, and curriculum in the link between education and labor market outcomes. He has looked at race and sex differences in employment and earnings. He is also studying the extended family as a source of support, the value of job seniority, the effectiveness of private schools, the effect of a school voucher program on public school students, black–white differences in wealth holdings, the determination of work hours, and econometric methods.

ALFRED BLUMSTEIN is the J. Erik Jonsson University Professor of Urban Systems and Operations Research and former dean (from 1986 to 1993) at the Heinz School of Public Policy and Management of Carnegie Mellon University. He also directs the National Consortium on Violence Research. He has had extensive experience in both research and policy with the criminal justice system. He served on the President's Crime Commission in 1966–1967 as director of its Task Force on Science and Technology. He has chaired National Academy of Sciences panels on research on deterrent and incapacitative effects, on sentencing, and on criminal careers. On the policy side, from 1979 to 1990, he chaired the Pennsylvania Commission on Crime and Delinquency, the state's criminal justice planning agency, and he served on the Pennsylvania Commission on Sentencing from 1986 to 1996. His degrees from Cornell University include a bachelor of engineering physics and a Ph.D. in operations research. He was elected to the National Academy of Engineering in 1998. Dr. Blumstein is a fellow of the American Society of Criminology, was the 1987 recipient of the society's Sutherland Award for "contributions to research," and was the president of the society in 1991–1992. His research over the past 20 years has covered many aspects of criminal justice phenomena and policy, including crime measurement, criminal careers, sentencing, deterrence and incapacitation, prison populations and racial disproportionality, demographic trends, juvenile violence, and drug enforcement policy.

LAWRENCE BOBO is professor of Afro-American studies and sociology and director of graduate studies in sociology at Harvard University. He was born in Nashville, Tennessee, and grew up in Los Angeles. He received a B.A. in sociology from Loyola Marymount University in 1979 and both the M.A. (1981) and Ph.D. (1984) in sociology from the University of Michigan. From 1984 through 1990 he was in the sociology department at the University of Wisconsin, Madison. From 1990 through spring of 1997 he was in UCLA's sociology department, where he also served, at various times, as associate chair, program director for survey research, and director of the Center for Research on Race, Politics and Society. His research interests constitute a fusion of race and ethnic relations (particularly the experience of African Americans in the post–World War II period), social psychology, public opinion and survey research methods: or, for lack of a more felicitous phrase: racial attitudes. He is co-author of *Racial Attitudes in America: Trends and Interpretations* (1987). He has been a fellow at the Center for Advanced Study in the Behavioral Sciences and a visiting scholar at the Russell Sage Foundation. He has served on the Board of Directors for the Social Science Research Council, the Executive Council's of the Interuniversity Consortium for Political and Social Research, the American Association for Public Opinion Research, the Association of Black Sociologists, the General Social Survey Board of Overseers, and the National Science Foundation Sociology Review Panel. He edited the Special Issue on Race of the journal *Public Opinion Quarterly* (Spring 1997).

CONSTANCE F. CITRO is a senior program officer for the Committee on National Statistics (CNSTAT). She is a former vice president and deputy director of Mathematica Policy Research, Inc., and was an American Statistical Association/National Science Foundation research fellow at the U.S. Census Bureau. For the committee, she has served as study director for numerous projects, including the Panel to Review the 2000 Census, the Panel on Estimates of Poverty for Small Geographic Areas, the Panel on Poverty and Family Assistance, the Panel to Evaluate the Survey of Income and Program Participation, the Panel to Evaluate Microsimulation Models for Social Welfare Programs, and the Panel on Decennial Census Methodology. Her research has focused on the quality and accessibility of large complex microdata files and analysis related to income and poverty measurement. She is a fellow of the American Statistical Association. She received a B.A. degree from the University of Rochester and M.A. and Ph.D. degrees in political science from Yale University.

MARILYN DABADY is a study director with CNSTAT. Her main areas of interest are interpersonal and intergroup relations; prejudice, stereotyping, and discrimination; and organizational behavior. She has conducted experimental research in social and cognitive psychology and has contributed to several National Academies reports on education and military recruitment. In addition to her duties as study director for this panel, Dr. Dabady also works with the Committee on the Youth Population and Military Recruitment. She received M.S. and Ph.D. degrees in psychology from Yale University.

DANELLE J. DESSAINT (project assistant) was a staff member of CNSTAT. Her projects included the Panel on Formula Allocations, State Children's Health Insurance Program, Elder Abuse, and Institutional Research Board studies. She has a B.A. in communications from Wingate University (Wingate, NC) and formerly worked as an editor at Tribune Media Services (Glens Falls, NY).

JOHN J. DONOHUE III, the William H. Neukom Professor of Law at Stanford Law School, is an economist/lawyer who has used large-scale statistical studies to estimate the impact of law and public policy in a wide range of areas, from civil rights and employment discrimination law to school funding and crime control. Professor Donohue is a Phi Beta Kappa graduate of Hamilton College and received a J.D. from Harvard and a Ph.D. in economics from Yale. In addition to his current appointment at Stanford, he has been on the faculty or visited at the law schools of Harvard, Yale, the University of Chicago, Northwestern, Cornell, and the University of Virginia and was a fellow at the Center for Advanced Studies in Behavioral Sciences in 2000–2001. He is the editor of the volume *Foundations of Employment Discrimination Law* (Foundation Press, 2nd ed., 2003), and the following are among his major articles on issues involving racial discrimination: "The Schooling of Southern Blacks: The Roles of Social Activism and Private Philanthropy, 1910–1960," *Quarterly Journal of Economics* (with James Heckman and Petra Todd, 2002, pp. 225–268); "The Impact of Race on Policing and Arrests," *Journal of Law and Economics* (with Steven Levitt, 2001, pp. 367–394); "Employment Discrimination Law in Perspective: Three Concepts of Equality," 92 *Michigan Law Review* 2583 (1994); "The Changing Nature of Employment Discrimination Litigation," 43 *Stanford Law Review* 983 (1991; with Peter Siegelman); and "Continuous versus Episodic Change: The Impact of Civil Rights Policy on the Economic Status of Blacks," 29 *Journal of Economic Literature* 1603 (December 1991; with James Heckman).

ROBERTO FERNANDEZ is the William F. Pounds Professor of Behavioral Policy Science at the Massachusetts Institute for Technology Sloan School of Management. His expertise lies in organizational process, social networks, hiring, turnover, and diversity. His research and teaching focuses on economic sociology, organizational behavior, social stratification, race, and ethnic relations. Among his current projects are networks and hiring and Internet-based recruitment. Recent published research includes *How Much Is That Network Worth? Social Capital in Employee Referral Networks* (with Emilio Castilla); *Social Capital: Theory and Research* (2001); *Social Capital at Work: Networks and Employment at a Phone Center* (with Emilio Castilla and Paul Moore); *American Journal of Sociology*, 2000; "Skill Biased Technological Change: Evidence from a Plant Retooling," *American Journal of Sociology* (2001).

STEPHEN E. FIENBERG is Maurice Falk University Professor of Statistics and Social Science in the Department of Statistics, the Center for Automated Learning and Discovery, and the Center for Computer and Communications Security at Carnegie Mellon University. He is a member of the

National Academy of Sciences and currently serves on the advisory committee of the National Research Council's Division of Behavioral and Social Sciences and Education. He is a past chair of CNSTAT and has served on several of its panels. He has published extensively on statistical methods for the analysis of categorical data and methods for disclosure limitation. His research interests include the use of statistics in public policy and the law, surveys and experiments, and the role of statistical methods in census taking.

SUSAN T. FISKE is professor of psychology at Princeton University, having taught on the faculties of the University of Massachusetts, Amherst, and Carnegie Mellon University. A 1978 Harvard Ph.D., she received an honorary doctorate from the Université Catholique de Louvain, Louvain-la-Neuve, Belgium, in 1995. She has authored over 150 journal articles and book chapters; she has edited 7 books and journal special issues. Her graduate text with Shelley Taylor, *Social Cognition* (1984; 2nd ed., 1991), defined the subfield of how people think about and make sense of other people. Her 2004 text, *Social Beings: A Core Motives Approach to Social Psychology*, describes people's most relevant evolutionary niche as social groups, with core motives (such as belonging) that enable people to adapt. Her research has focused on how people choose between category-based (stereotypic) and individuating impressions of other people, as a function of power and interdependence. Her current research shows that social structure predicts distinct kinds of bias against different groups in society, some more disrespected and others more disliked. Her expert testimony in discrimination cases includes one cited by the U.S. Supreme Court in a 1989 landmark case on gender bias. In 1998, she also testified before President Clinton's Race Initiative Advisory Board. Dr. Fiske won the 1991 American Psychological Association Award for Distinguished Contributions to Psychology in the Public Interest, Early Career, in part for the expert testimony. She also won, with Glick, the 1995 Allport Intergroup Relations Award from the Society for the Psychological Study of Social Issues for work on ambivalent sexism. Among other elected offices, Dr. Fiske was president of the American Psychological Society for 2002–2003. She edited, with Daniel Gilbert and Gardner Lindzey, the *Handbook of Social Psychology* (4th ed., 1998) and with Daniel Schacter and Carolyn Zahn-Waxler, the *Annual Review of Psychology* (Vols. 51–60, 2000–2009). She has served on the boards of Scientific Affairs for the American Psychological Association, the American Psychological Society, Annual Reviews Inc., the Social Science Research Council, and the Common School in Amherst.

MARISA A. GERSTEIN is a research assistant with CNSTAT. She has worked on a diverse number of projects, including panels on elder mistreatment, nonmarket accounts, research and development statistics, and the 2000 and 2010 censuses. She is a coeditor of *Statistical Issues in Allocating Funds by Formula*, the final report issued by the Panel on Formula Allocations. She graduated from New College of Florida with a B.A. in sociology.

GLENN C. LOURY is currently university professor, professor of economics, and director of the Institute on Race and Social Division at Boston University. Previously he taught economics at Harvard University, Northwestern University, and the University of Michigan. He earned a B.A. in mathematics at Northwestern University and holds a Ph.D. in economics from the Massachusetts Institute of Technology. Professor Loury has made scholarly contributions to the fields of welfare economics, game theory, industrial organization, natural resource economics, and the economics of income distribution. He has been a scholar in residence at Oxford University, Tel Aviv University, the University of Stockholm, the Delhi School of Economics, the Institute for the Human Sciences in Vienna, and the Institute for Advanced Study at Princeton. Professor Loury has received a Guggenheim Fellowship to support his work. He is a fellow of the American Academy of Arts and Sciences and the Econometric Society and was elected vice president of the American Economics Association for 1997. His most recent book, *The Anatomy of Racial Inequality*, appeared in February 2002 from the Harvard University Press.

SAMUEL R. LUCAS is currently an associate professor of sociology at the University of California, Berkeley. His research and teaching interests lie in social stratification, sociology of education, methods, and statistics. Professor Lucas has served on the Editorial Board of Sociology of Education, as a consulting editor for the *American Journal of Sociology*, and on the sociology advisory panel of the National Science Foundation and as a member of the National Academy of Sciences Committee on the Representation of Minority Students in Special Education. He is currently serving on the Technical Review Panel for the Education Longitudinal Study of 2002. He has published in *Social Forces*, *Sociology of Education*, and the *American Journal of Sociology* and coauthored *Inequality by Design: Cracking the Bell Curve Myth* with five colleagues in the Sociology Department at Berkeley, which received a Gustavus Myers Center Award for the Study of Human Rights in North America in 1997. His book on tracking, titled *Tracking Inequality: Stratification and Mobility in American High Schools*, received the Willard Waller Award in 2000 for the most outstanding book in the sociology of education for 1997, 1998, and 1999. He is completing a book on the effects of race and sex discrimination in the United States.

DOUGLAS S. MASSEY is professor of sociology and public policy at the

Woodrow Wilson School of Public and International Affairs at Princeton University. He formerly served on the faculties of the University of Chicago and the University of Pennsylvania. He is the coauthor of numerous books and articles on racial segregation, discrimination, and immigration, including the award-winning book *American Apartheid: Segregation and the Making of the Underclass*. He is a member of the National Academy of Sciences and the American Academy of Arts and Sciences.

JANET L. NORWOOD is a counselor and senior fellow at the New York Conference Board, where she chairs the Advisory Committee on the Leading Indicators. She served as U.S. commissioner of labor statistics from 1979 to 1992 and then was a senior fellow at the Urban Institute until 1999. She chaired the Advisory Council on Unemployment Compensation from 1993 to 1996 and from 1992 to 1999 was a member of CNSTAT. She has been a member of the Division of Engineering and Physical Sciences and has served as chair or as a member of several committee panels. She chairs the Advisory Committee for the Bureau of Transportation Statistics and serves as a member of the Board of Scientific Counselors to the National Center for Health Statistics. She holds a B.A. from Douglass College and an M.A. and a Ph.D. from Tufts University and has received honorary LL.D.'s from Rutgers, Harvard, Carnegie Mellon, and Florida International universities. She is a fellow and past president of the American Statistical Association, a member and past vice president of the International Statistical Institute, an honorary fellow of the Royal Statistical Society, and a fellow of the National Academy of Public Administration and the National Association of Business Economists.

JOHN E. ROLPH is professor of statistics in the Department of Information and Operations Management at the Marshall School of Business, University of Southern California. He also holds faculty appointments in the mathematics department and in the law school at the University of Southern California. He previously was on the research staff of the RAND Corporation. He has also held faculty positions at University College London, Columbia University, the RAND Graduate School for Policy Studies, and the Health Policy Center of RAND/University of California at Los Angeles. He received A.B. and Ph.D. degrees in statistics from the University of California at Berkeley. He is a fellow of the American Association for the Advancement of Science, the American Statistical Association, and the Institute of Mathematical Statistics and is an elected member of the International Statistical Institute. He is currently chair of CNSTAT of the National Academies and is a member of the CNSTAT Panel on Operational Test Design and Evaluation of the Interim Armored Vehicle. His research interests include empirical Bayes methods and the application of statistics to legal and public policy issues.

Index

A

B

C

D

E

F

G

H

I

J

L

M

N

O

P

R

T

U

V

W